Spring Boot企业级
开发入门与实战 （IntelliJ IDEA·微课视频版）

丁明浩 刘仲会 ◎ 主编

清华大学出版社
北京

内 容 简 介

随着移动互联网的发展，对 Web 开发的需求日益上升。Spring Boot 作为 Web 开发领域中的利器，无论是单体应用，还是用于面向服务架构或者微服务架构，都有不错的表现。本书面向准备在 Web 开发领域深入学习的读者，详细介绍了 Spring Boot 2.3 应用开发的相关知识。从功能点出发，每章都是不同的 Spring Boot 应用之旅。

全书共 8 章。第 1、2 章介绍了 Spring Boot 2.3 开发基础，包括环境搭建、依赖引入和基础 Spring Boot 应用构建。第 3~6 章是 Spring Boot 的融合阶段，介绍了利用 Spring Boot 搭建 Web 项目、操作数据库、使用缓存、整合安全框架、结合消息服务等，这些都是日常开发中的常用内容，读者经过该阶段的学习可以初步运用 Spring Boot 进行敏捷开发。第 7、8 章是 Spring Boot 的实战阶段，读者经过该阶段的学习可以更加熟练地运用 Spring Boot，从而掌握实际项目的开发技能。

本书的特点是示例代码丰富，实用性和系统性较强，读者可以直接还原书中的示例。本书适合作为高等院校计算机及相关专业的教材或教学参考书，也可作为相关开发人员的自学教材或参考手册。

本书封面贴有清华大学出版社防伪标签，无标签者不得销售。
版权所有，侵权必究。举报: 010-62782989, beiqinquan@tup.tsinghua.edu.cn。

图书在版编目(CIP)数据

Spring Boot 企业级开发入门与实战: IntelliJ IDEA: 微课视频版/丁明浩,刘仲会主编. —北京: 清华大学出版社,2023.9
(清华科技大讲堂丛书)
ISBN 978-7-302-62966-5

Ⅰ.①S… Ⅱ.①丁… ②刘… Ⅲ.①JAVA 语言－程序设计 Ⅳ.①TP312.8

中国国家版本馆 CIP 数据核字(2023)第 039899 号

策划编辑: 魏江江
责任编辑: 王冰飞
封面设计: 刘　键
责任校对: 时翠兰
责任印制: 沈　露

出版发行: 清华大学出版社
网　　址: http://www.tup.com.cn, http://www.wqbook.com
地　　址: 北京清华大学学研大厦 A 座　　邮　编: 100084
社 总 机: 010-83470000　　邮　购: 010-62786544
投稿与读者服务: 010-62776969, c-service@tup.tsinghua.edu.cn
质量反馈: 010-62772015, zhiliang@tup.tsinghua.edu.cn
课件下载: http://www.tup.com.cn, 010-83470236
印 装 者: 三河市人民印务有限公司
经　　销: 全国新华书店
开　　本: 185mm×260mm　　印　张: 19.5　　字　数: 472 千字
版　　次: 2023 年 9 月第 1 版　　印　次: 2023 年 9 月第 1 次印刷
印　　数: 1~1500
定　　价: 59.80 元

产品编号: 099210-01

前言

党的二十大报告指出：教育、科技、人才是全面建设社会主义现代化国家的基础性、战略性支撑。必须坚持科技是第一生产力、人才是第一资源、创新是第一动力，深入实施科教兴国战略、人才强国战略、创新驱动发展战略，这三大战略共同服务于创新型国家的建设。高等教育与经济社会发展紧密相连，对促进就业创业、助力经济社会发展、增进人民福祉具有重要意义。

在项目开发中，微服务是极其常见的开发架构。以前，公司多使用单体项目部署微服务，无论是打包还是运行都耗时耗力，每次需要创建新应用、构建项目，配置 Spring 时都十分麻烦。当前，许多公司已经将 Spring Boot 作为企业应用程序开发的主要框架，对于采用微服务架构的 REST API 尤其如此。Spring Boot 是 Spring 社区中的顶级项目，在整个生态中如同基石一样。无论是想结合模板引擎实现一个单体应用、支撑前端项目的 RESTful 服务，还是基于 Spring Cloud 开发一套微服务，这些都离不开 Spring Boot。

本书从实际应用出发，理论结合实例，深入浅出地对 Spring Boot 开发进行讲解。实战内容将贯穿全书，指导读者通过动手实践，从一行语句、一个方法到整个项目，完整地理解 Spring Boot 开发的流程，从而获得并提升 Web 应用开发的能力。

本书共 8 章，各章内容如下。

第 1 章 Spring Boot 入门，讲解 Spring Boot 开发所需的基本概念，主要包括工具选择、环境搭建、项目构建和基础的开发流程。

第 2 章 Spring Boot 核心配置与注解，主要介绍 Spring Boot 自动化配置、配置文件属性值注入和多环境配置。

第 3 章 Spring Boot 视图技术，介绍构建基于 Spring Boot 的单体应用所需掌握的基本知识，主要包括实现页面国际化、异常统一处理、文件上传等内容，同时还讲解了如何构建 RESTful Web 服务。

第 4 章 Spring Boot 数据访问，着重探讨数据持久化技术，依次讲解简单易懂的 JDBC、整合轻量级框架 MyBatis 和方便快捷的 ORM 解决方案 JPA。整个过程层层递进，帮助读者理解不同场景下数据库在 Spring Boot 中的调用方法。

第 5 章 Spring Boot 安全管理，重点讲解应用程序的安全性，Spring Boot 的安全可以通过整合 Spring Security 框架来实现。

第 6 章 Spring Boot 消息服务，主要介绍 RabbitMQ 的安装与使用、Spring Boot 集成 RabbitMQ、利用 RabbitMQ 实现不同类型的消息服务。

第 7 章基于 Spring Boot＋Shiro＋Vue 开发的前后端分离学生信息管理项目整合实战——后端开发，介绍用 Spring Boot 作为开发环境，整合 Shiro、Jwt、MyBatis 和 Redis 开发后端的学生信息管理项目的实际案例。重点讲解了 Spring Boot 对 Shiro 框架及 Redis 的整合等，对前后端分离项目的关键技术进行了较为详细的讲解和设计使用。

第 8 章基于 Spring Boot＋Shiro＋Vue 开发的前后端分离学生信息管理项目整合实战——前端开发，介绍以 VSCode 作为开发环境，整合 Node.js、Element-plus、Axios、Vuex 等技术，实现基于 Vue 开发前端框架的学生信息管理项目的实际案例，并对如何实现跨域请求、角色动态路由渲染、Axios 请求和响应拦截器的封装等进行了详细的设计和实现。

本书项目实战开发环境为 Windows 10，开发工具使用 IntelliJ IDEA 2018.2，JDK 使用 1.8 版本，Tomcat 使用 8.0 版本，Spring Boot 使用 2.6.2 版本。在学习本书之前，读者需要掌握 J2SE 基础知识和 Java Web 的相关技术，如 Spring、HTML、Tomcat、MyBatis 等技术。此外，读者需要掌握主流数据库基本知识，如 MySQL 等，掌握其基本的 SQL 语法和常用数据库的安装。

为便于教学，本书提供丰富的配套资源，包括教学大纲、教学课件、电子教案、教学进度表、习题答案、程序源码和微课视频。

资源下载提示

数据文件：扫描目录上方的二维码下载。

微课视频：扫描封底的文泉云盘防盗码，再扫描书中相应章节的视频讲解二维码，可以在线学习。

本书能够顺利出版，首先要感谢清华大学出版社给笔者一次和大家分享技术、交流学习的机会，感谢各位编辑在本书出版过程中的辛勤付出。张居彦副教授和刘仲会老师在本书的编写过程中付出了很多辛勤的汗水，在此一并表示衷心的感谢。

由于编者水平及写作时间有限，书中难免会有不妥之处，敬请各位读者批评、指正。

编　者
2023 年 6 月

目录

资源下载

第 1 章　Spring Boot 入门 ····· 1

1.1　Spring Boot 概述 ····· 1
1.1.1　Spring Boot 和 MVC 架构的对比 ····· 1
1.1.2　Spring Boot 简介 ····· 3
1.1.3　Spring Boot 的特征 ····· 3

1.2　Spring Boot 开发准备 ····· 5
1.2.1　什么是 Maven ····· 5
1.2.2　配置开发环境 ····· 7
1.2.3　使用 Maven 方式构建 Spring Boot 项目 ····· 11
1.2.4　使用 Spring Initializr 方式构建 Spring Boot 项目 ····· 15

1.3　Spring Boot 文件目录 ····· 19
1.3.1　Java 类文件 ····· 19
1.3.2　资源文件 ····· 20
1.3.3　测试类文件 ····· 20
1.3.4　pom.xml 文件 ····· 20

1.4　热部署 ····· 21

1.5　单元测试 ····· 24
1.5.1　单元测试模板 ····· 25
1.5.2　测试 Service 层 ····· 25
1.5.3　测试 Controller 层 ····· 29

1.6　打包与部署 ····· 31
1.6.1　以 JAR 包方式运行 ····· 32
1.6.2　以 WAR 包方式运行 ····· 34

本章小结 ····· 36
习题 ····· 36

第 2 章　Spring Boot 核心配置与注解 ····· 38

2.1　自动化配置 ····· 38
2.1.1　@SpringBootApplication ····· 39
2.1.2　SpringApplication ····· 41

2.2 全局配置 ··· 45
2.3 自定义配置 ··· 47
  2.3.1 注入自定义属性到字段中 ·· 47
  2.3.2 注入自定义属性到对象中 ·· 48
  2.3.3 注入自定义配置文件 ·· 50
  2.3.4 自动扫描配置类 ·· 53
2.4 多环境配置 ··· 54
  2.4.1 使用 Profile 进行多环境配置 ·· 54
  2.4.2 使用@Profile 进行多环境配置 ··· 55
本章小结 ··· 57
习题 ·· 57

第 3 章 Spring Boot 视图技术 ·· 59

3.1 创建静态 Web 页面 ·· 59
3.2 Spring Boot 对 JSP 的支持 ··· 61
3.3 Thymeleaf 的基本语法 ·· 63
  3.3.1 变量表达式 ··· 64
  3.3.2 自定义变量 ··· 65
  3.3.3 方法 ··· 66
  3.3.4 字面值 ·· 67
  3.3.5 拼接 ··· 68
  3.3.6 运算 ··· 68
  3.3.7 循环 ··· 70
  3.3.8 逻辑判断 ·· 71
  3.3.9 分支控制 switch ··· 72
  3.3.10 Thymeleaf 模板片段 ··· 73
3.4 实现基于 Thymeleaf 的 Web 应用 ·· 75
3.5 Spring Boot 中的页面国际化实现 ··· 78
3.6 Spring Boot 集成 Spring MVC ··· 85
  3.6.1 配置自定义拦截器 Interceptor ·· 85
  3.6.2 跳转指定页面 ·· 87
3.7 Spring Boot 处理 JSON 数据 ·· 87
3.8 Spring Boot 实现 RESTful 风格的 Web 应用 ·· 90
3.9 Spring Boot 文件上传和下载 ·· 93
  3.9.1 文件上传 ·· 94
  3.9.2 文件下载 ·· 96
3.10 Spring Boot 的异常统一处理 ·· 99
  3.10.1 自定义 error 页面 ··· 99
  3.10.2 @ExceptionHandler 注解 ·· 101

3.10.3 @ControllerAdvice 注解 ·· 103
本章小结 ·· 106
习题 ·· 106

第 4 章　Spring Boot 数据访问 ·· 109

4.1 Spring Boot 整合 JDBC ·· 109
　　4.1.1 Spring Data 简介 ·· 109
　　4.1.2 整合 JDBC Template ·· 110
　　4.1.3 数据库连接池 Druid ·· 113
4.2 Spring Boot 整合 MyBatis ··· 115
　　4.2.1 使用配置文件的方式整合 MyBatis ··· 117
　　4.2.2 使用注解的方式整合 MyBatis ··· 119
4.3 Spring Boot 整合 JPA ·· 122
　　4.3.1 Spring Data JPA 简介 ·· 123
　　4.3.2 简单条件查询 ·· 124
　　4.3.3 关联查询 ·· 129
　　4.3.4 @Query 和@Modifying 注解 ··· 138
　　4.3.5 排序和分页查询 ··· 142
4.4 数据缓存 Cache ··· 146
本章小结 ·· 152
习题 ·· 152

第 5 章　Spring Boot 安全管理 ·· 154

5.1 Spring Security 简介 ··· 154
　　5.1.1 什么是 Spring Security ·· 154
　　5.1.2 为什么要使用 Spring Security ·· 155
　　5.1.3 Spring Security 的核心类 ·· 155
5.2 安全管理效果测试 ··· 157
5.3 自定义用户认证 ·· 159
　　5.3.1 内存身份认证 ·· 160
　　5.3.2 JDBC 身份认证 ·· 161
　　5.3.3 UserDetailsService 身份认证 ·· 163
5.4 自定义用户授权管理 ·· 166
　　5.4.1 授权基本流程 ·· 166
　　5.4.2 自定义登录页面 ··· 169
　　5.4.3 权限控制和注销 ··· 173
　　5.4.4 "记住我"及首页定制 ·· 177
本章小结 ·· 180
习题 ·· 180

第 6 章　Spring Boot 消息服务 ··· 183

- 6.1　消息服务概述 ·· 183
- 6.2　Exchange 策略 ··· 186
- 6.3　消息的各种机制 ··· 188
- 6.4　安装 RabbitMQ ·· 189
  - 6.4.1　什么是 RabbitMQ ·· 189
  - 6.4.2　RabbitMQ 安装过程 ··· 190
- 6.5　Spring Boot 整合 RabbitMQ ··· 195
  - 6.5.1　简单消息的发送和接收 📹 ··· 196
  - 6.5.2　发布订阅模型 📹 ·· 199
  - 6.5.3　会员注册模型 📹 ·· 203
- 本章小结 ··· 208
- 习题 ··· 208

第 7 章　基于 Spring Boot＋Shiro＋Vue 开发的前后端分离学生信息管理项目整合实战——后端开发 ··· 211

- 7.1　开发思路整合 ·· 211
- 7.2　系统设计 ·· 211
  - 7.2.1　系统功能需求分析 ··· 212
  - 7.2.2　系统模块划分 ·· 212
  - 7.2.3　数据库设计 ··· 214
- 7.3　后端系统环境搭建 ··· 221
  - 7.3.1　使用 Maven 组件为项目添加依赖 JAR 包 ······································· 221
  - 7.3.2　项目的目录结构 ··· 224
  - 7.3.3　项目的配置文件 ··· 225
  - 7.3.4　项目的配置类 ·· 226
- 7.4　Apache Shiro 的工作机制和配置类设计 ·· 230
  - 7.4.1　Shiro 的工作机制 ··· 230
  - 7.4.2　Shiro 配置类 ShiroConfig 设计 ·· 232
- 7.5　基于 Shiro 框架的用户登录设计 ·· 237
  - 7.5.1　用户登录的实体类设计 ·· 237
  - 7.5.2　用户登录设计 ·· 239
  - 7.5.3　项目的启动类 DemoApplication ·· 242
  - 7.5.4　项目的启动测试 ··· 242
- 7.6　Apache Shiro 认证授权安全框架设计 ··· 243
  - 7.6.1　Shiro 的认证授权工作流程 ·· 243
  - 7.6.2　findByUsername 请求的组件设计 ·· 244
  - 7.6.3　JWTFilter 类对 token 值的过滤设计 ··· 245

7.6.4 CustomRealm 类对当前登录用户身份验证设计 …… 247
7.6.5 CustomRealm 类对当前登录用户授权设计 …… 249
7.6.6 GlobalExceptionHandler 全局异常捕获设计 …… 250
7.7 Apache Shiro 认证授权测试用例 …… 251
7.7.1 findByUsername 请求成功用例 …… 251
7.7.2 findByUsername 请求身份认证失败用例 …… 252
7.7.3 findByUsername 请求授权认证失败用例 …… 253
7.7.4 用户授权 Redis 缓存管理测试 …… 254
7.8 后端接口设计 …… 256
7.8.1 findRoutesByRole 接口设计 …… 256
7.8.2 findByName 接口设计 …… 261
7.8.3 commitgraphbyuser 接口设计 …… 263
7.8.4 用户退出登录 logout 接口设计 …… 264

第 8 章 基于 Spring Boot＋Shiro＋Vue 开发的前后端分离学生信息管理项目整合实战——前端开发 …… 266

8.1 开发思路整合 …… 266
8.2 前端系统环境搭建 …… 267
8.2.1 Vue 框架介绍 …… 267
8.2.2 前端环境搭建 …… 267
8.2.3 创建 Vue 3.0 项目 …… 270
8.2.4 项目目录结构 …… 271
8.3 前端项目的配置文件 …… 272
8.3.1 package.json …… 272
8.3.2 App.vue …… 274
8.3.3 main.js …… 274
8.3.4 vue.config.js …… 275
8.4 前端用户登录模块设计 …… 277
8.4.1 用户登录页面 Login.vue 设计 …… 277
8.4.2 用户登录拦截器 api.js 设计 …… 281
8.4.3 用户请求 Controller 接口的 axios.js 设计 …… 282
8.4.4 获取动态路由 menus.js 设计 …… 284
8.4.5 用户登录成功页面显示 …… 288
8.5 前端用户个人信息管理模块设计 …… 290
8.5.1 用户信息页面 UserInfo.vue 设计 …… 290
8.5.2 更新密码的隐式表单设计 …… 292
8.5.3 提交电子签名表单设计 …… 294
8.6 前端用户注销登录模块设计 …… 296

参考文献 …… 299

第1章

Spring Boot入门

本章学习目标
- 了解 Spring Boot 概述。
- 熟悉 Spring Boot 应用的开发环境。
- 掌握 Spring Boot 应用的构建。
- 掌握 Spring Boot 应用程序单元测试。
- 掌握 Spring Boot 打包与部署。

根据党的二十大报告,在发展思路上,既要着力破解难题、补齐短板,又要考虑巩固和厚植原有优势。"不畏浮云遮望眼,只缘身在最高层"。坚持系统性、整体性、协同性等都是辩证思维方法的集中体现。

Spring Boot 是当下最为流行的 Java 端开发框架,该框架由 Spring 开源组织提供,使用该框架可以解决传统项目中混乱的 Maven 依赖管理问题,同时可以基于 Maven 快速进行项目的打包和发布。

通过本章的学习,读者可以深刻理解马克思主义辩证法:整体与局部的关系。

1.1 Spring Boot 概述

与传统的 Spring MVC 框架相比,通过使用基于 Spring Boot 的开发模式,可以简化搭建框架时配置文件的数量,从而提升系统的可维护性。在 Spring Boot 框架中,还可以更方便地引入如安全和负载均衡等方面的组件。更具体地说,Spring Boot 框架是微服务的基础,在这个框架中可以引入 Spring Cloud 的诸多组件,从而搭建基于微服务的系统。

1.1.1 Spring Boot 和 MVC 架构的对比

在 Java 项目开发中,MVC 已经成为一种深入人心的设计模式,几乎所有正规的项目中都会使用 MVC 设计模式。采用 MVC 设计模式可以有效地实现显示层、控制层、业务层、数据层的结构分离,如图 1-1 所示。

虽然 MVC 开发具有良好的可扩展性,但是在实际的开发过程中,许多开发者依然会感受到如下的问题。

(1) 采用原生 Java 程序实现 MVC 设计模式时,一旦整体项目设计不到位,就会存在大量的重复代码,并且项目维护困难。

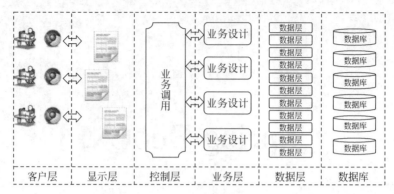

图 1-1　MVC 设计模式

（2）为了简化 MVC 各层的开发，可以引用大量的第三方开发框架，如 Spring、Hibernate、MyBatis、JPA、SpringSecurity 等，但这些框架都需要在 Spring 中实现整合，其结果就是会存在大量的配置文件。

（3）当使用一些第三方的服务组件（如 RabbitMQ、JavaMail 等）时，需要编写大量重复的配置文件，而且还需要根据环境定义不同的 profile（如 dev、product 等）。

（4）使用 Maven 作为构建工具时，需要配置大量的依赖关系，且程序需要被打包为 WAR 文件并部署到应用服务器上才可以执行。

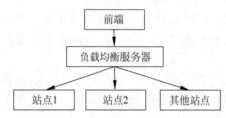

图 1-2　基于分布式的在线购物网站的架构

（5）RESTful 作为接口技术，其应用越来越广泛。如果使用 Spring 来搭建 RESTful 服务，则需要引入大量的 Maven 依赖库，并且需要编写许多的配置文件。

在实际项目里，为了应对高并发的访问请求，往往会进行分布式部署，如图 1-2 所示。在这种分布式架构，从前端页面发到后端的大量请求会被负载均衡服务器分发到不同的服务器处理，还可以把数据库做成集群，用多台数据库服务器分担高并发的压力。

从实际效果上来看，如果采用图 1-2 所示的分布式架构，用多台业务处理服务器和数据库服务器，确实能满足高并发的需求。不过根据实践经验，上述架构一般会存在如下问题。

（1）各功能模块之间的调用关系会比较复杂，即耦合度比较高。一个模块的修改往往会影响到其他多个模块，即代码比较难维护。

（2）由于在具体的每台机器上是集中式部署，因此稳定性不强，往往一个问题会导致整个系统崩溃。即使采用基于分布式的主从冗余等措施，这个问题也无法得到根本解决。

（3）可扩展性不强。假设当前的并发量是每秒 100 次请求，用两台服务器即可满足需求。但当业务量增加，并发量上升到每秒 2000

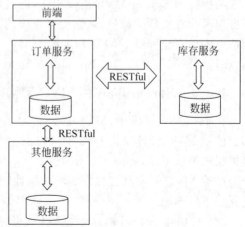

图 1-3　微服务的体系结构

次请求后,就需要再次扩展服务器,从而造成不便。

与上述架构相比,微服务(Microservice)的体系结构如图 1-3 所示。由图 1-3 可知,微服务模块之间一般会通过 RESTful 格式的请求进行通信,换句话说,模块间的耦合度比较低,这样就便于在任何模块里变更业务需求。

此外,每个模块都具有自己的数据库,也就是说,每个模块都能独立运行,整个系统的扩展性比较强,能用比较小的代价来扩展新的功能模块。

1.1.2 Spring Boot 简介

微服务是体系架构,或者说是模块的组织形式。通俗地讲,如果用"微服务架构"的方式组装业务模块,那么整个系统就能具有"高扩展性"和"模块间低耦合度"的特性。

注意,微服务是一个抽象的概念,它有不同的实现方式,而基于 Spring Boot 是当前比较流行的一种实现微服务的方式。

由于 Spring 具备控制反转(Inversion of Control,IoC)的特性,因此通过 Spring 开发出来的模块,它们之间的耦合度非常低,这与微服务的要求非常相似。在之前 Spring 版本的基础上,Pivotal 团队提供了一套全新的 Spring Boot 框架。在这套框架里,开发者可以嵌入 Web 服务器(如 Tomcat),无须把项目文件打包并部署到 Web 服务器上,而且 Spring Boot 还具备自动配置的功能。更为便利的是,通过定义配置文件,开发者还能"自动监控"基于 Spring Boot 框架模块的各项运行时的性能指标。总之,可以这样理解,Spring Boot 框架是 Spring 框架的升级版,通过之后基于代码的叙述,可以更加详细地体会 Spring Boot 框架的优势。

Spring Boot 是 Spring 开发框架提供的一种扩展支持,其主要目的是希望通过简单的配置实现开发框架的整合,使开发者的注意力可以完全放在程序业务功能的实现上,其核心在于通过"零配置"的方式来实现快速且简单的开发。另外,Spring Boot 通过集成大量的框架,使得依赖包的版本冲突和引用的不稳定性等问题得到了很好的解决。图 1-4 为 Spring Boot 当前的开发版本。

2.6.1 CURRENT GA	Reference Doc.	API Doc.
2.6.2-SNAPSHOT SNAPSHOT	Reference Doc.	API Doc.
2.5.8-SNAPSHOT SNAPSHOT	Reference Doc.	API Doc.
2.5.7 GA	Reference Doc.	API Doc.
2.4.13 GA	Reference Doc.	API Doc.
2.3.12.RELEASE GA	Reference Doc.	API Doc.

图 1-4 Spring Boot 支持的开发版本

1.1.3 Spring Boot 的特征

相较于传统的 Spring 框架,Spring Boot 框架具有以下特征。

1. 可快速构建独立的 Spring 应用

Spring Boot 是一个依靠大量注解实现自动化配置的全新框架。在构建 Spring 应用时,只需要添加相应的场景依赖,Spring Boot 就会根据添加的场景依赖自动进行配置,在无须额外手动添加配置的情况下快速构建出一个独立的 Spring 应用。

Spring Boot 具有多种快速构建项目的方式,几种常用的形式如下:

（1）使用 Eclipse（MyEclipse）创建项目。可以利用创建 Maven 项目的方式创建 Spring Boot 项目。当然，如果在 Eclipse 中安装了 Spring Tools，就可以直接创建 Spring Starter Project。

（2）使用 IntelliJ IDEA 创建项目。可以利用创建 Spring Initializr 的方式创建 Spring Boot 项目，在后续章节会详细介绍这种方式的过程。

（3）使用 Spring Tool Suite 创建项目。可以直接新建 Spring Starter Project 项目，过程类似 Eclipse 创建 Spring Boot 项目。

（4）使用官方文档创建项目。Spring 官方文档提供了一种在线生成 Spring Boot 项目的方式。首先访问 Spring 官方快速构建地址，然后在这个页面上选择对应版本、构建工具等，填写完成后单击 Generate Project 按钮，即可在本地下载一个 Spring Boot 项目的压缩包。

2．遵循习惯优于配置的原则

Spring Boot 的配置都在 application.properties 文件中，但是并不意味着在 Spring Boot 应用中就必须包含该文件。application.properties 配置文件包含大量的配置项，而大多数配置项都有其默认值，使用默认值即可完成配置。这类配置称为"自动化配置"。

3．直接嵌入 Tomcat 等服务器（无须部署 WAR 文件）

传统的 Spring 应用部署时，通常会将应用打包成 WAR 形式并部署到 Tomcat 服务器中。Spring Boot 框架内嵌了 Tomcat、Jetty 等服务器，而且可以自动将项目打包，并在项目运行时部署到服务器中。

4．提供了"开箱即用"的 Spring 插件

在 Spring Boot 项目构建过程中，无须准备各种独立的 JAR 文件，只需在构建项目时根据开发场景需求选择对应的依赖启动器 starter 即可。在引入的依赖启动器 starter 内部已经包含了对应开发场景所需的依赖，并会自动下载和拉取相关 JAR 包。例如，在 Web 开发时，只需在构建项目时选择对应的 Web 场景依赖启动器 spring-boot-starter-web，Spring Boot 项目便会自动导入 spring-webmvc、spring-web、spring-boot-starter-tomcat 等子依赖，并自动下载和获取 Web 开发需要的相关 JAR 包。

5．自动化配置 Spring 和第三方库

Spring Boot 充分考虑与传统 Spring 框架及其他第三方库融合的场景，在提供了各种场景依赖启动器的基础上，内部还默认提供了各种自动化配置类。在使用 Spring Boot 开发项目时，一旦引入了某个场景的依赖启动器，Spring Boot 内部提供的默认自动化配置类就会生效，开发者无须手动在配置文件中进行相关配置，从而极大减少了开发人员的工作量，提高了程序的开发效率。

6．多环境配置

在项目开发过程中，项目不同的角色会使用不同的环境，比如开发人员会使用开发环境、测试人员会使用测试环境、性能测试会使用性能测试环境、项目开发完成之后会把项目部署到线上环境等。不同的环境往往会连接不同的 MySQL 数据库、Redis 缓存、MQ 消息中间件等。环境之间相互独立与隔离才不会相互影响，隔离的环境便于部署和提高工作效率。假设项目需要 3 个环境，即开发环境、测试环境和性能测试环境，复制项目配置文件 application.properties，分别取名为 application-dev.properties、application-test.properties 和 application-perform.properties，依次作为开发环境、测试环境和性能测试环境。多环境的配置文件开发完成后，在配置文件 application.properties 中添加配置激活选项，具体代码如下所示。

（1）激活开发环境配置：

spring.profiles.active = dev

（2）激活测试环境配置：

spring.profiles.active = test

（3）激活性能测试环境的配置：

spring.profiles.active = perforn

7．极少的代码生成和 XML 配置

Spring Boot 框架内部已经实现了与 Spring 框架及其他常用第三方库的整合连接，并提供了默认最优化的整合配置，使用时基本上不需要额外生成配置代码和 XML 配置文件。在需要自定义配置的情况下，Spring Boot 更加提倡使用 Java config 替换传统的 XML 配置方式，这样更加方便查看和管理。

1.2 Spring Boot 开发准备

在开始学习 Spring Boot 之前，需要准备好开发环境。本节将以 Windows 操作系统为例，介绍如何安装 JDK、IntelliJ IDEA 和 Apache Maven。如果所使用的计算机上已经安装了 JDK、IntelliJ IDEA 或 Apache Maven，可以略过本节内容。

1.2.1 什么是 Maven

1. Maven 简介

在用 Eclipse 开发项目时，一定会引入支持特定功能的 JAR 包。例如，由图 1-5 可知，该项目需要引入支持 MySQL 的 JAR 包。

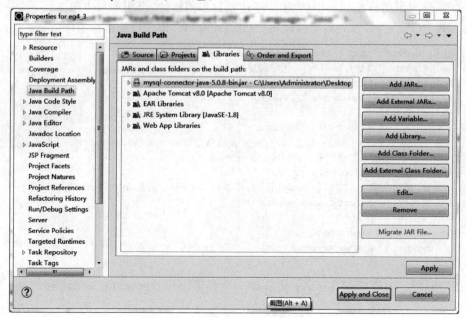

图 1-5　在项目里引入 JAR 包的示意图

如图1-5所示，支持MySQL的JAR包是放在本地路径里的，这样在本地运行时自然是没问题的，但要把这个项目发布到服务器上就会出现问题，这是因为该项目的.classpath文件已经指定MySQL的JAR包在本地C盘下的某个路径中。

```
<classpathentry kind="lib" path="C:/Users/Administrator/Desktop/mysql-connector-java-5.0.8-bin.jar"/>
<classpathentry kind="output" path="build/classes"/>
```

一旦发布到服务器上，项目会根据.classpath的配置从C盘下的路径去寻找，但事实上是不可能有这样的路径和JAR包的。

也可以通过在.classpath里指定相对路径来解决这个问题，在下面的代码里指定本项目将引入"本项目路径/WebRoot/lib"目录里的JAR包。

```
<classpathentry kind="lib" path="webroot/mysql-connector-java-5.0.8-bin.jar"/>
<classpathentry kind="output" path="build/classes"/>
```

发布到服务器时，由于会把整个项目路径里的文件都上传，因此不会出错，但这样依然会带来不便。例如，在服务器上部署了5个项目，它们都会用到MySQL支持包，这样就不得不把这个JAR包上传5次。再扩展一下，如果5个项目里会用到20个相同的JAR包，那么就需要进行多次复制。如果要升级其中的一个JAR包，那么就需要进行多次重复的复制粘贴动作。

期望中的工作模式应该是，有一个"仓库"统一放置所有的JAR包，在开发项目时，可以通过配置文件引入必要的包，而不是把包复制到项目里。这就是Maven的做法。

通俗地讲，Maven是一套Eclipse插件，它的核心价值是能理顺项目间的依赖关系。具体来讲，Maven能通过其中的pom.xml配置文件来统一管理本项目所要用到的JAR包，在项目里引入Maven后，开发者就不必手动添加JAR包，这样也能避免一系列问题。

2. POM文件的构成

POM（Project Object Model，项目对象模型）是Maven工程的基本工作单元，它是一个XML文件，包含了项目的基本信息，用于描述项目如何构建、声明项目依赖等。

在创建POM之前，首先需要描述项目组（groupId），即项目的唯一ID。

```
<project xmlns="http://maven.apache.org/POM/4.0.0"
    xmlns:xsi="http://www.w3.org/2001/XMLSchema-instance"
    xsi:schemaLocation="http://maven.apache.org/POM/4.0.0
    http://maven.apache.org/xsd/maven-4.0.0.xsd">
    <!-- 模型版本 -->
    <modelVersion>4.0.0</modelVersion>
    <!-- 公司或者组织的唯一标志，配置时生成的路径由此生成，如it.com.boot。Maven会将该项目打包成的JAR包放入本地路径：/it/com/boot -->
    <groupId>it.com.boot</groupId>
    <!-- 项目的唯一ID，一个groupId下面可能有多个项目，根据artifactId进行区分 -->
    <artifactId>project</artifactId>
    <!-- 版本号 -->
    <version>1.0</version>
    <packaging>jar</packaging>
</project>
```

所有POM文件都需要project元素和3个必需字段：groupId、artifactId、version，如表1-1所示。

表1-1 POM文件的构成

节 点	描 述
project	工程的根标签
modelVersion	模型版本需要设置为4.0
groupId	工程组的标识。它在一个组织或者项目中通常是唯一的
artifactId	工程的标识。它通常是工程的名称。groupId 和 artifactId 一起定义了 artifact 在仓库中的位置
version	工程的版本号。在 artifact 的仓库中,它用来区分不同的版本。例如,it.com.boot:project:1.0、it.com.boot:project:1.1
packaging	项目打包的类型。可以是 JAR、WAR、RAR、EAR、POM,默认是 JAR

3. dependencies 和 dependency

dependencies 和 dependency 的关系为前者包含后者。如前所述,Maven 的一个重要作用就是统一管理 JAR 包,为了一个项目可以构建或运行,项目中不可避免地会依赖很多其他的 JAR 包,这些依赖在 Maven 中被称为 dependency。

例如,项目中用到了 MyBatis,那么 POM 可以添加如下配置:

```
<dependencies>
    <dependency>
        <groupId>org.mybatis</groupId>
        <artifactId>mybatis</artifactId>
        <version>3.2.5</version>
    </dependency>
</dependencies>
```

按 Ctrl+S 组合键保存,Eclipse 就会自动在远程仓库里下载 MyBatis 的 JAR 包到本地仓库,通过 groupId、artifactId、version 唯一标识一个 Maven 项目。有了3个必需字段才可以到远程仓库下载,如果本地仓库里已经有该 JAR 包,则不会到远程仓库下载。

1.2.2 配置开发环境

1. 安装 JDK

JDK(Java SE Development Kit)建议使用1.8及以上的版本,可以根据 Windows 操作系统的配置选择合适的 JDK1.8 安装包。软件下载完成后,双击下载软件,出现如图1-6所示的安装界面,一直单击"下一步"按钮即可完成安装。这里将 JDK 安装在 C:\Java\jdk1.8.0_152 目录下。

安装完成后,需要设置环境变量,具体步骤如下:

(1) 在计算机桌面上右击"我的电脑",选择"属性"→"高级系统设置"→"环境变量"选项,在"系统变量(S)"中单击"新建"按钮,弹出新建环境变量的窗口,如图1-7所示。

(2) 在"变量名"和"变量值"中分别输入 JAVA_HOME 和 C:\Java\jdk1.8.0_152,单击"确定"按钮。

(3) JAVA_HOME 配置完成后,将%JAVA_HOME%\bin 加入到"系统变量"的 path 中。然后,打开命令行窗口,输入命令 java-version,出现如图1-8所示的提示,即表示安装成功。

图 1-6　JDK 安装界面

图 1-7　新建环境变量的窗口

图 1-8　JDK 安装成功的命令行窗口

2. 安装 IntelliJ IDEA

在 IntelliJ IDEA 的官方网站上可以免费下载 IDEA。IDEA 下载完成后，运行安装程序，按照提示安装即可。本书使用 IntelliJ IDEA 2018.2.4 版本，其软件界面如图 1-9 所示。

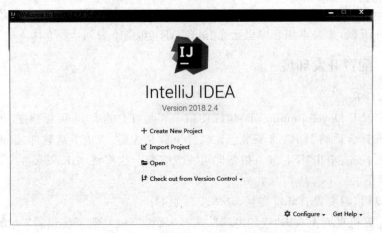

图 1-9　IntelliJ IDEA 软件界面

3. 安装 Maven

Maven 是一个比较常用的项目管理工具，同时提供了出色的应用程序构建能力。通常可以通过几行命令构建一个简单的应用程序。在 Apache Maven 官方网站可以下载最新版本的 Maven，本书使用的 Maven 版本为 apache-maven-3.6.0。

下载完成后解压缩即可，然后将 Maven 的安装路径 G:\apache-maven\bin 加入到 Windows 的环境变量 path 中。安装完成后，在命令行窗口执行命令 mvn -v，如果输出如

图1-10所示的页面,则表示Maven安装成功。

图1-10　Maven安装成功的命令行窗口

Maven环境变量至此配置完成。在默认情况下,Maven下载JAR包可能会有一些慢,可以修改为国内华为、阿里云等下载地址。用户可以根据需求自行修改代码如下所示。

```xml
<?xml version="1.0" encoding="UTF-8"?>
<settings>
<!-- 需要修改为自己的Maven本地仓库地址 -->
<localRepository>G:/apache-maven/MavenRepository</localRepository>
   <servers>
    <server>
        <id>huaweicloud</id>
        <username>anonymous</username>
        <password>devcloud</password>
    </server>
   </servers>
   <mirrors>
    <mirror>
        <id>huaweicloud</id>
        <mirrorOf>*</mirrorOf>
        <url>https://mirrors.huaweicloud.com/repository/maven/</url>
    </mirror>
   </mirrors>
   <profiles>
    <!-- 修改Maven默认的JDK版本 -->
    <profile>
     <id>JDK-1.8</id>
     <activation>
         <activeByDefault>true</activeByDefault>
         <jdk>1.8</jdk>
     </activation>
     <properties>
        <maven.compiler.source>1.8</maven.compiler.source>
        <maven.compiler.target>1.8</maven.compiler.target>
        <maven.compiler.compilerVersion>1.8</maven.compiler.compilerVersion>
     </properties>
    </profile>
   </profiles>
</settings>
```

4. IntelliJ IDEA配置JDK和Maven

1) JDK初始化设置

在IntelliJ IDEA界面中,选择Configure→Project Defaults→Project Structure选项,进

入 Project Structure 设置页面。在设置界面左侧选择 Project Settings→Project 选项，在打开的右侧页面中进行 JDK 初始化设置，如图 1-11 所示。

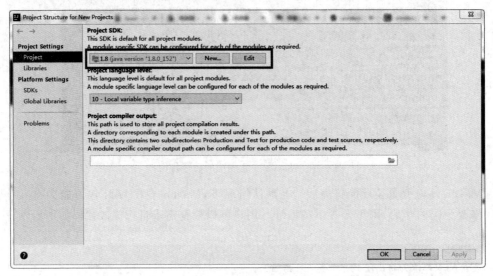

图 1-11　JDK 初始化设置

在图 1-11 所示界面中，可以通过单击右侧页面的 New 按钮选择自定义安装的 JDK 路径，设置完成后，依次单击 Apply→OK 按钮完成 JDK 的初始化配置。

2) Maven 初始化设置

在 IntelliJ IDEA 界面中，选择 Configure→Project Defaults→Settings 选项，在弹出的对话框中找到 Maven 选项，分别把 Maven home directory、User settings file、Local repository 设置为自己 Maven 的相关目录，如图 1-12 所示。设置完成后，依次单击 Apply→OK 按钮完成 Maven 的初始化配置。

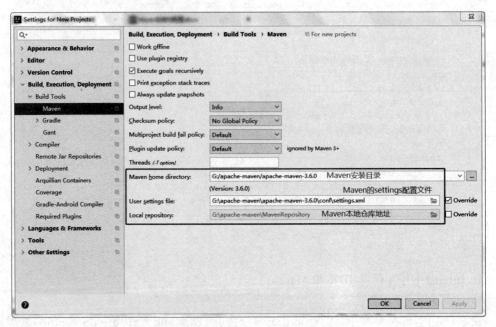

图 1-12　Maven 初始化设置

1.2.3 使用 Maven 方式构建 Spring Boot 项目

Apache Maven 是目前比较流行的自动化构建工具。需要声明的是，IDEA 创建的 Spring Boot 项目，可以不使用 Tomcat 作为服务器，Tomcat 已经内嵌到 Spring Boot 的 JAR 包中，不用配置 Tomcat 服务器也能运行 JavaWeb 项目（这是因为 Spring Boot 内置了 Tomcat）。

1) 新建项目

在 IntelliJ IDEA 界面中，依次单击 File 菜单→New 按钮→Project 选项，在弹出的 New Project 对话框中新建 Maven 工程，如图 1-13 所示。

视频讲解

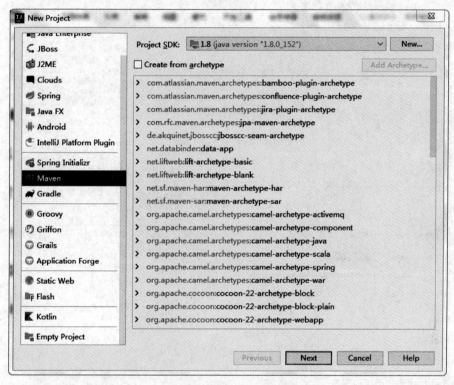

图 1-13 新建 Maven 工程

在图 1-13 所示界面中，左侧是可以选择创建的项目类型，包括 Spring 项目、Android 项目、Spring Initializr 项目和 Maven 项目等，右侧是不同类型项目对应的设置界面。左侧选择 Maven 选项，右侧选择当前项目的 JDK，单击 Next 按钮进入 Maven 项目创建界面。

2) 输入项目基本信息

在 Group 栏输入组织名，在 Artifact 栏输入项目名，其他选择默认，直接单击 Next 按钮，如图 1-14 所示。

3) 确认项目信息

输入项目名称，选择项目保存位置。在图 1-15 所示界面中，Project name 用于指定项目名称，在上一步中定义的 ArtifactId 会默认作为项目名；Project location 用于指定项目的存储路径。这里使用上一步设置的 springboot0101 作为项目名称，单击右侧的…按钮可以修改存放路径。项目名称和存放路径设置完成后，单击 Finish 按钮完成项目的创建。

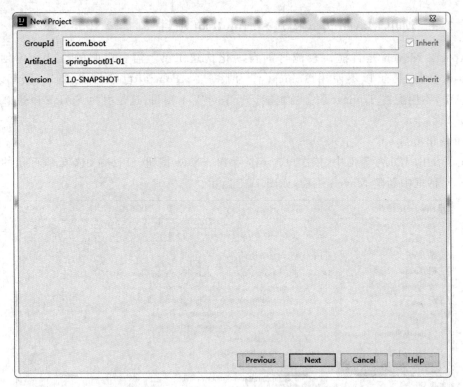

图 1-14　输入 Maven 项目基本信息

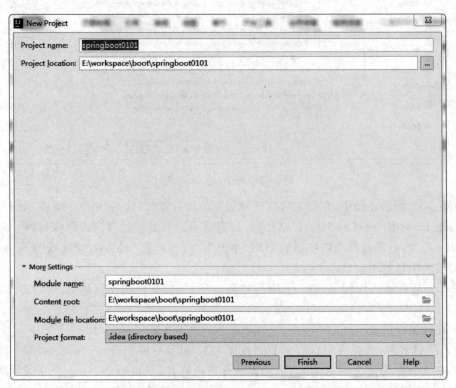

图 1-15　输入项目名称和项目保存位置

4）添加 Spring Boot 相关依赖

打开 springboot0101 项目下的 pom.xml 文件，在该 pom.xml 文件中添加构建 Spring Boot 项目和 Web 场景开发对应的依赖，示例代码如下。

```xml
<parent>
    <groupId>org.springframework.boot</groupId>
    <artifactId>spring-boot-starter-parent</artifactId>
    <version>2.1.3.RELEASE</version>
</parent>
<dependencies>
    <dependency>
        <groupId>org.springframework.boot</groupId>
        <artifactId>spring-boot-starter-web</artifactId>
    </dependency>
</dependencies>
```

在 pom.xml 文件中添加依赖，这里使用 parent 标签定义了标签的依赖版本。spring-boot-starter-parent 是一个特殊的 starter，通过使用 Dependency Management 进行项目依赖的版本管理，这样常用的包依赖就不再需要设置版本号。同时，该 starter 还提供了下列默认配置：

- 编码格式默认是 UTF-8；
- Java 版本默认是 1.8；
- 自动化的资源过滤；
- 自动化的插件配置；
- 定义构建打包的配置。

spring-boot-starter-web 依赖包可以快速构建一个可执行的 Web 应用，不需要进行其他的 Web 配置。该依赖中包含了很多常用的依赖包，如 spring-web、spring-webmvc 等，默认使用嵌入式 Tomcat 发送服务请求。

修改 POM 文件后，窗口右下角会提示导入依赖包，单击 Import Changes，如图 1-16 所示。

此时会进行 Maven 依赖的下载，如图 1-17 所示。

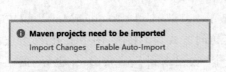

图 1-16　导入 Maven 依赖包

图 1-17　Maven 依赖的下载

5）编写启动类

在 src/main/java 目录上右击，创建一个 it 包，并在包中创建 HelloApplication 主类，如图 1-18 所示。

在图 1-18 中，第 6 行代码中的 @SpringBootApplication 注解是 Spring Boot 框架的核心注解，该注解用于表明 HelloApplication 类是 Spring Boot 项目的主程序启动类。第 10 行代码调用 SpringApplication.run() 方法启动主程序类。

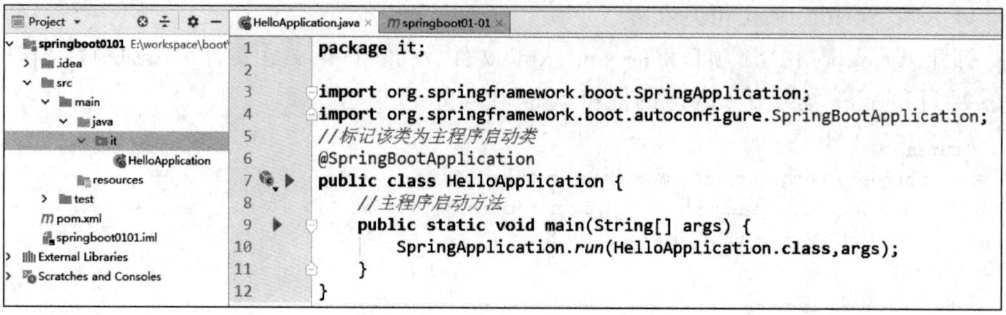

图 1-18 创建包和主类

6）编写 Controller 类

接下来创建一个 Spring MVC 中的控制器 HelloController。需要注意的是，该控制类必须创建在启动类或其目录包的下面才行，否则 Spring Boot 启动类无法扫描到 HelloController.java 文件。添加测试映射器代码如图 1-19 所示。

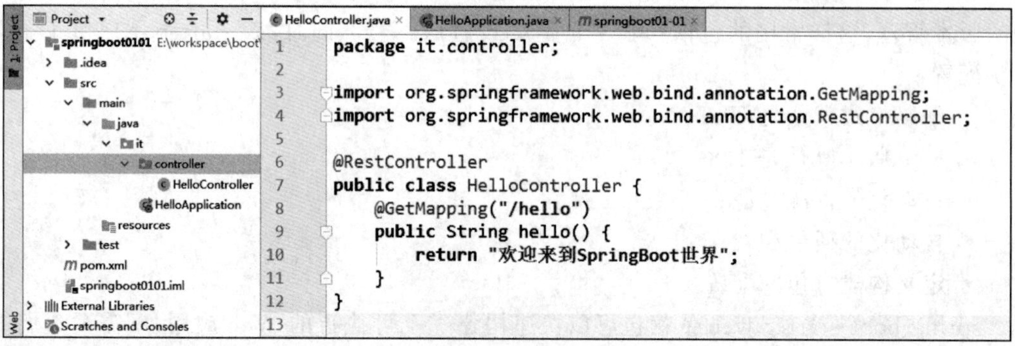

图 1-19 编写 Controller 类

在图 1-19 中，请求处理控制类 HelloController 和请求处理方法 hello()都使用了注解，其中：

（1）@RestController 注解是一个组合注解，等同于@Controller 和@ResponseBody 两个注解结合使用的效果。该注解的主要作用是将当前类作为控制层的组件添加到 Spring 容器中，同时该类的方法无法返回 JSP 页面，而会返回 JSON 字符串。

（2）@GetMapping 注解等同于@RequestMapping(method＝RequestMethod.GET)注解，其主要作用是设置方法的访问路径并限定其访问方式为 GET。图中 hello()方法的请求处理路径为"/hello"，并且方法的返回值是一个"欢迎来到 Spring Boot 世界"的字符串对象。

7）启动项目

在第一次创建的主程序类中，右击 Run HelloApplication 或者直接单击工具栏中的运行按钮，内置的 Tomcat 服务器也会同时启动，如图 1-20 所示。

此时在浏览器地址栏中输入 localhost:8080/hello，就可以访问先前创建的第一个控制器类中的方法，并输出内容到浏览器中显示，如图 1-21 所示。

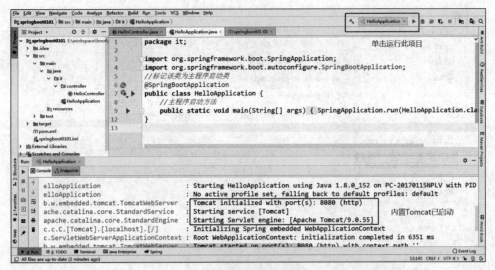

图 1-20　启动项目

图 1-21　Web 应用运行结果

1.2.4　使用 Spring Initializr 方式构建 Spring Boot 项目

视频讲解

由于使用的开发工具是 IDEA，开发者可以借助 Spring Initializr 简单快速地创建一个 Spring Boot 项目。这种方式不但可以生成完整的目录结构，而且可以生成一个默认的主程序，以节省开发时间。

1) 新建项目

打开 New Project 界面后，在左侧选择 Spring Initializr，如图 1-22 所示。选择后右边会有两个选项，第一项是选择 JDK 版本，由于本书均采用 Spring Boot 2.6.1 版本，最低支持 JDK 1.8 版本，因此选择 JDK 1.8。第二项为 Initializr Service URL，用来查询 Spring Boot 的当前版本和组件的网站。这两个选项配置完成后，单击 Next 按钮进入下一个步骤。

2) 配置 Project Metadata 信息

在 IDEA 创建新项目时，要根据项目情况设置项目的元数据（metadata），设置项目元数据的界面如图 1-23 所示。

如图 1-23 所示，在所创建项目 Group 文本框中输入 it.com.boot，在 Artifact 文本框中输入 springboot01-02，在所创建项目的管理工具类型 Type 中选择 Maven Project。开发语言 Language 选择 Java，打包方式 Packaging 选择 Jar，开发工具 Java 的版本 Java Version 选择 8（也称为 1.8）。所创建项目的版本 Version 保留自动生成的 0.0.1-SNAPSHOT，项目名称 Name 保留自动生成的 springboot01-02。项目描述 Description 可以修改为"使用

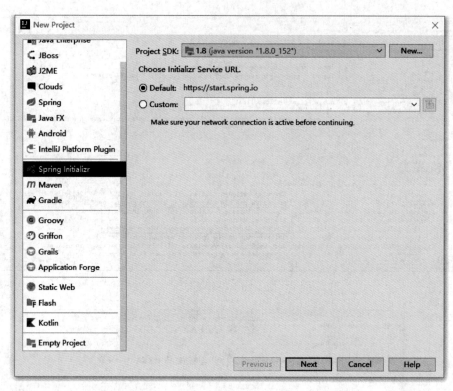

图 1-22　新建 Spring Initializr 工程

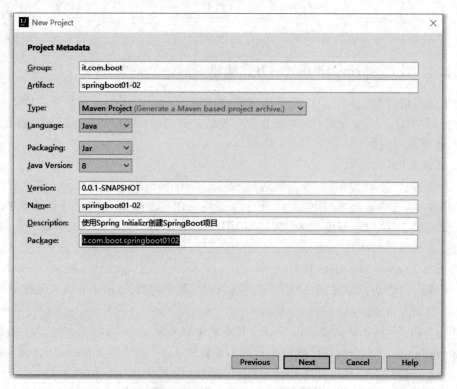

图 1-23　输入项目基本信息

Spring Initializr 创建 SpringBoot 项目",所创建项目默认的包名 Package 可以修改为"it.com.boot.springboot0102"。

3)添加组件

填写完项目的元数据后,单击 Next 按钮就可以进入选择项目依赖(Dependencies)的界面。如图 1-24 所示,IDEA 自动选择了 Spring Boot 的最新版本(本例中的 2.6.1 版),也可以手动选择所需要的版本,再手动为所创建的项目(springboot01-02)选择 Web 依赖。选择完 Web 依赖后,IDEA 就可以帮助开发者完成 Web 项目的初始化工作。创建项目时,也可以不选择任何依赖,而在文件 pom.xml 中添加所需要的依赖。

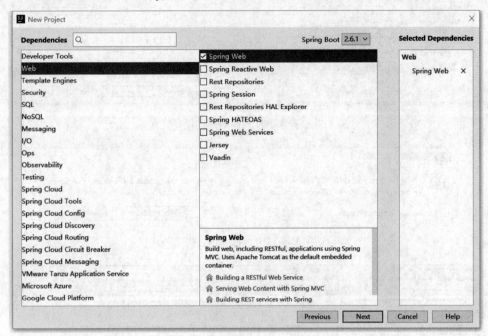

图 1-24　选择版本和组件

4)确认项目信息

在图 1-24 中,单击 Next 按钮后,进入项目名称(Project Name)和项目位置(Project Location)的显示页面,可以直接保留由图 1-25 生成的项目名称和位置默认值,也可以根据需要直接修改项目名称和项目位置。然后单击 Finish 按钮,就可以进入项目界面。

如果开发工具弹出如图 1-26 所示的对话框,则提示已经打开其他的项目,新创建的项目是否要在新的窗口打开。单击 New Window 按钮表示在新的窗口打开项目,单击 This Window 按钮表示在当前窗口打开项目,这里选择 New Window 按钮即可。

5)编写 Controller 类

在自动生成的目录和文件的基础上,在 it.com.boot.springboot0102 包下新建 controller 子包,然后在 controller 子包中创建 HelloWorldController,如图 1-27 所示。

6)启动项目

运行入口类 Springboot0102Application,成功启动自带的内置 Tomcat。在浏览器地址栏中输入 localhost:8080/hello 后,浏览器中显示 Web 应用运行结果如图 1-28 所示。

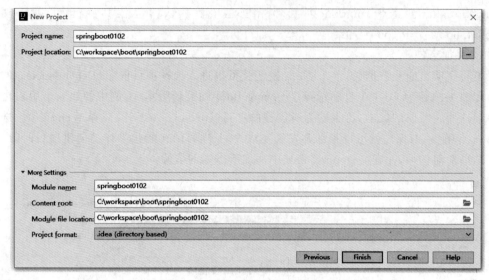

图 1-25　输入项目名称和项目保存位置

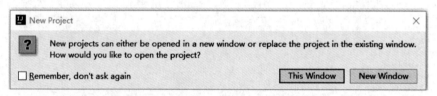

图 1-26　新窗口打开项目提示

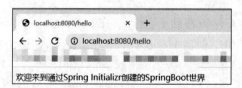

图 1-27　编写 Controller 类

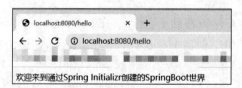

图 1-28　Web 应用运行结果

1.3 Spring Boot 文件目录

创建 Spring Boot 项目后，会产生一个工程目录，该工程目录存放了工程项目的各种文件。对于 Spring Boot 开发人员来说，了解该工程目录非常必要。从图 1-29 中可以看到，Spring Boot 文件大致分为以下 4 部分。

（1）Java 类文件。

（2）资源文件。

（3）测试类文件。

（4）POM 文件。

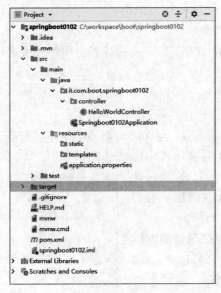

图 1-29 IntelliJ IDEA 项目工程介绍

1.3.1 Java 类文件

src/main/java 目录用于放置 Java 类文件。由于这是一个新建的项目，因此目前只有一个 Springboot0102Application 类，如图 1-30 所示。这个类是 Spring Boot 应用的主程序，其中@SpringBootApplication 注解用来说明这是 Spring Boot 应用的启动类，包含自动配置、包扫描等功能；main 方法是启动应用的入口方法，以命令行或者插件等任何方式启动，都会调用这个方法。

```
Springboot0102Application.java

package it.com.boot.springboot0102;

import ...

@SpringBootApplication
public class Springboot0102Application {

    public static void main(String[] args) {
        SpringApplication.run(Springboot0102Application.class, args);
    }

}
```

图 1-30 Springboot0102Application 类

1.3.2 资源文件

1. 配置文件

src/main/resources 目录主要用于放置 Spring Boot 应用的配置文件。新建项目时会默认创建一个 application.properties(默认为空文件),也可以将 application.properties 文件修改为 application.yml 文件,用缩进结构的键值对来进行配置。同时,配置文件可以进行一些应用需要的配置,如端口号等。

2. 静态资源

src/main/resources/static 目录主要放置应用的静态资源文件,如 HTML、JavaScript、图片等。

3. 模板文件

src/main/resources/templates 目录主要放置应用的模板文件,如使用 Thymeleaf 后的 Thymeleaf 模板文件等。

1.3.3 测试类文件

src/test/java 目录用于放置 Spring Boot 测试类文件,默认会根据项目名称创建一个测试类,如图 1-31 所示。打开该类可以发现@SpringBootTest 注解用于表明这是一个 Spring Boot 测试类。

(1) @SpringBootTest:此注解能够测试 SpringApplication。因为 Spring Boot 程序的入口是 SpringApplication,基本上所有配置都会通过入口类进行加载,而该注解可以引用入口类的配置。

(2) @Test:JUnit 单元测试的注解,注解在方法上表示一个测试方法。

图 1-31 Springboot0102ApplicationTests 类

当执行 Springboot0102ApplicationTests.java 中的 contextLoads 方法时,可以看到控制台和执行入口类中的 SpringApplication.run()方法打印的信息是一致的。由此可知,@SpringBootTest 引入了入口类的配置。

1.3.4 pom.xml 文件

Spring Boot 项目下的 pom.xml 文件主要用来存放依赖信息,具体代码如下所示。

```
<parent>
    <groupId>org.springframework.boot</groupId>
    <artifactId>spring-boot-starter-parent</artifactId>
    <version>2.6.1</version>
    <relativePath/><!-- lookup parent from repository -->
</parent>

<dependencies>
```

```xml
<dependency>
    <groupId>org.springframework.boot</groupId>
    <artifactId>spring-boot-starter-web</artifactId>
</dependency>

<dependency>
    <groupId>org.springframework.boot</groupId>
    <artifactId>spring-boot-starter-test</artifactId>
    <scope>test</scope>
</dependency>
</dependencies>

<build>
    <plugins>
        <plugin>
            <groupId>org.springframework.boot</groupId>
            <artifactId>spring-boot-maven-plugin</artifactId>
        </plugin>
    </plugins>
</build>
```

其中 spring-boot-starter-parent 和 spring-boot-starter-web 已经介绍过，spring-boot-starter-test 依赖与测试相关，只要引入它，就会把所有与测试相关的包全部引入。spring-boot-maven-plugin 是一个 Maven 插件，能够以 Maven 的方式为应用提供 Spring Boot 的支持，即为 Spring Boot 应用提供执行 Maven 操作的可能，并能够将 Spring Boot 应用打包为可执行的 JAR 或 WAR 文件。

pom.xml 文件中的依赖都有一个 scope 属性，其属性值分别如下：

- compile：默认值，表示被依赖项目需要参与当前项目的编译和后续的测试、运行周期，是一个比较强的依赖，打包时通常需要包含进去。
- test：依赖项目仅仅参与测试相关的工作，包括测试代码的编译和执行，不会被打包，如 junit。
- runtime：表示被依赖项目无须参与项目的编译，不过后期的测试和运行周期需要其参与。与 compile 相比，只是跳过了编译过程而已。例如 JDBC 驱动，适用运行和测试阶段。
- provided：打包时可以不用包含进去，其他设施会提供。该依赖理论上可以参与编译、测试、运行等周期，相当于 compile，但是打包阶段做了 exclude 操作。
- system：从参与度来说，与 provided 相同，不过被依赖项不会从 Maven 仓库下载，而是从本地文件系统获取，需要添加 systemPath 的属性来定义路径。

1.4 热部署

在项目开发过程中，常常会改动页面数据或者修改数据结构，为了显示改动效果，需要重启应用。这个过程其实就是重新编译生成了新的 Class 文件，该文件记录着与代码等对应的各种信息，将被虚拟机的 ClassLoader 加载。

热部署正是利用了这个特点，它监听到 Class 文件改动时，就会创建一个新的

ClassLoader 进行加载该文件，不需要重新启动应用，最终也能显示改动效果。

Spring Boot 提供了热部署方案，大大提高了开发效率。频繁地重启项目会浪费很多时间，有了热部署后，就不用担心修改代码会重启项目了。

下面通过实例讲解如何使用 spring-boot-devtools 进行热部署。

1. 创建 Spring Boot Web 应用项目 springboot0103

2. 添加 DevTools 依赖

Spring Boot 提供了一个名为 spring-boot-devtools 的模块来使应用支持热部署，从而提高开发者的开发效率，无须手动重启 Spring Boot 应用。在开发 Web 项目过程中，虽然改动项目重启总是报错，但 Spring Boot 对调试支持很好，修改后可以实时生效。热部署需要添加的配置如下。

```xml
<!-- Spring Boot 热部署配置 -->
<dependency>
    <groupId>org.springframework.boot</groupId>
    <artifactId>spring-boot-devtools</artifactId>
    <optional>true</optional>
</dependency>
```

3. 设置 IDEA 热部署工具

由于热部署是监听 Class 文件的变化，因此它自身不会主动编译 Java 文件。在 Java 文件改动时，需要将其自动编译成 Class 文件，然后热部署工具创造的新的类加载器才会加载改变后的 Class 文件。所以，如果使用 IDEA 开发工具的话，切记要把自动编译打开。

（1）开启 IDEA 的自动编译（静态）。

在 File 菜单中选择 Settings→Compiler 选项，在右侧勾选 Build Project automatically 选项，将项目设置为自动编译，依次单击 Apply→OK 按钮保存设置，如图 1-32 所示。

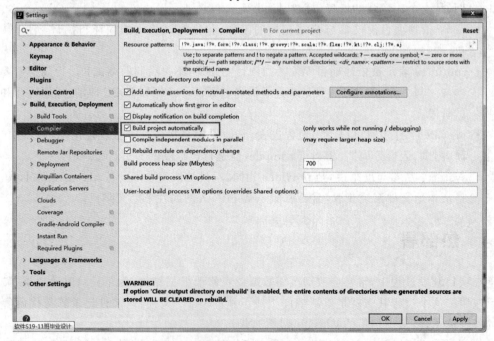

图 1-32　IDEA Compiler 配置

(2) 开启 IDEA 的自动编译(动态)。

在 Maintenance 界面按 Ctrl＋Alt＋Shift＋W 组合键，选择 Registry 选项，如图 1-33 所示。勾选 compiler.automake.allow.when.app.running 选项，将程序运行方式设置为自动编译，单击 Close 按钮完成配置。Registry 界面如图 1-34 所示。

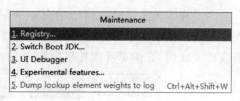

图 1-33 Maintenance 界面

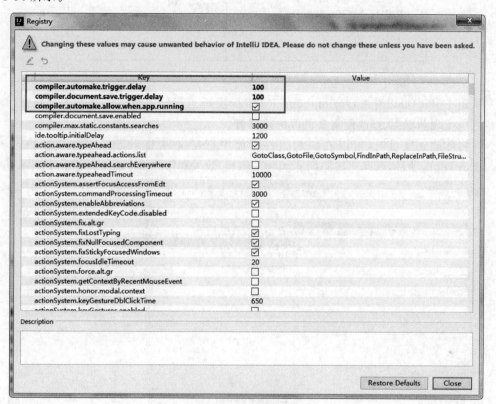

图 1-34 Registry 界面

在 Registry 界面中选择 compile.document.save.trigger.delay 选项，主要是针对静态文件如 JS、CSS 的更新，将延迟时间减少后，直接按 F5 键刷新页面就能看到效果。

4. 热部署测试

启动项目，在浏览器地址栏中输入 http://localhost:8080/hello，如图 1-35 所示。

图 1-35 访问/hello 接口

为了在不重新启动项目的情况下测试热部署是否配置成功，在 HelloWorldController 类中新增一个请求并保存，查看控制台信息会发现项目能够自动构建和编译，说明项目热部署生效。此时，在浏览器地址栏中输入 http://localhost:8080/say，如图 1-36 所示。

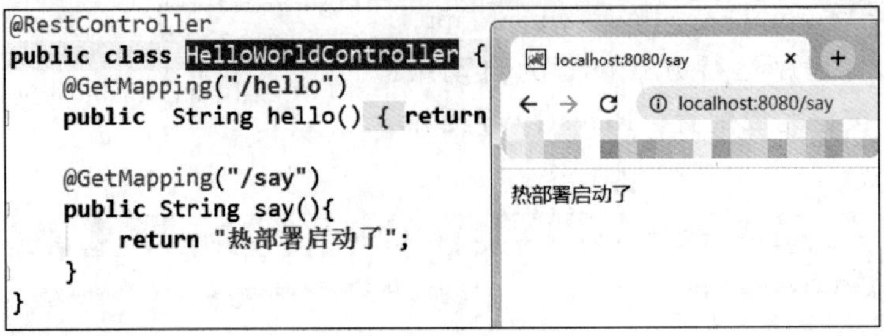

图 1-36 访问/say 接口

从图 1-36 中可以看出，spring-boot-devtools 实现了代码修改后的热部署。同样，该依赖也可以实现新增类、修改配置文件等的热部署。

在实际建立项目时，可以在添加组件界面中直接选择 Developer Tools 中的 Spring Boot DevTools 项目，以此完成添加 devtools 依赖，效果和第二步一样，如图 1-37 所示。

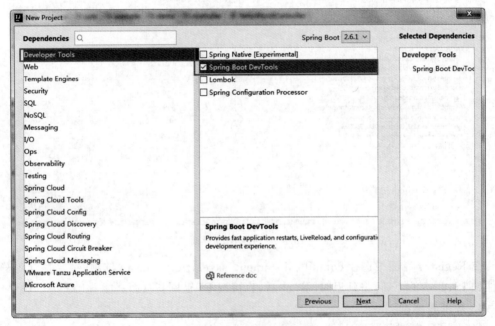

图 1-37 勾选热部署依赖

1.5 单元测试

在 Spring 项目中，对于 Controller、Service、DAO 各层都需要建立单元测试项。对应不同的分层，可以使用 JUnit 和 Mock 等不同的方式。然而有些情况需要启动 Spring 容器来测试业务逻辑在容器内能否正常运行，Spring Boot 针对此情况对项目的单元测试提供了很好的支持方式。

Spring Boot 为测试提供了一个名为 spring-boot-starter-test 的启动器，只要使用它就能引入 Spring Boot 测试模块，还能引入如 JUnit、AssertJ、Hamcrest 等有用的类库，具体如下所示。

- JUnit：Java 应用程序单元测试标准类库。
- Spring Test & Spring Boot Test：Spring Boot 应用程序功能集成化测试支持。
- AssertJ：轻量级的断言类库。
- Hamcrest：对象匹配器类库。
- Mockito：Java Mock 测试框架。
- JSONassert：用于 JSON 的断言库。
- JsonPath：JSON 操作类库。

1.5.1 单元测试模板

1. 添加依赖

在项目中使用 Spring Boot Test 支持时，只需要在 pom.xml 引入如下配置即可。

```xml
<dependency>
    <groupId>org.springframework.boot</groupId>
    <artifactId>spring-boot-starter-test</artifactId>
    <scope>test</scope>
</dependency>
```

2. 编写测试类

测试类代码如下：

```java
@RunWith(SpringRunner.class)
@SpringBootTest
public class Springboot0102ApplicationTests {
    @Test
    public void contextLoads() {
        //调用测试方法
    }
}
```

@SpringBootTest 和@Test 注解已经介绍过，在此介绍@RunWith 注解。@RunWith 注解是 JUnit 标准的一个注解，目的是告诉 JUnit 框架不要使用内置的方式进行单元测试，而应使用@RunWith 指明的类进行单元测试，所有的 Spring 单元测试总是使用 SpringRunner.class。

1.5.2 测试 Service 层

JUnit 4.4 引入了 Hamcrest 框架，Hamcrest 提供了一套匹配符 Matcher，这些匹配符更接近自然语言，可读性高且更加灵活。

以前 JUnit 提供了很多的 assertion 语句，如 assertEquals、assertFalse、assertNull 等，现在 JUnit 4.4 的一条 assertThat 语句即可替代所有的 assertion 语句。这样可以在所有的单元测试中只使用一个断言方法，使得编写测试用例变得简单，代码风格变得统一，测试代码也更容易维护。

assertThat 语法如下：

assertThat(T actual, Matcher<T> matcher);
assertThat(String reason, T actual, Matcher<T> matcher);

其中 actual 为需要测试的变量，matcher 为使用 Hamcrest 的匹配符来表达变量 actual 期望值的声明。

下面列举了一些常用的 Matcher 匹配符表达式，如表 1-2 所示。

表 1-2 常用的 Matcher 匹配符表达式

名称	示例	说明
allOf	assertThat（testedNumber，allOf（greaterThan(8)，lessThan(16)））；	如果接下来的所有条件都成立，则测试通过，相当于"与"（&&）
anyOf	assertThat(testedNumber, anyOf(greaterThan(16), lessThan(8)));	如果接下来的所有条件中有一个成立，则测试通过，相当于"或"（||）
is	assertThat（testedString，is（"developerWorks"））；	如果前面待测的 object 等于后面给出的 object，则测试通过
containsString	assertThat（testedString，containsString（"developerWorks"））；	如果测试的字符串 testedString 包含了字符串 developerWorks，则测试通过
startsWith	assertThat(testedString, startsWith("developerWorks"));	如果测试的字符串 testedString 以子字符串 developerWorks 开始，则测试通过
greaterThan	assertThat(testedNumber, greaterThan(16.0));	如果所测试的数值 testedNumber 大于 16.0，则测试通过
greaterThanOrEqualTo	assertThat(testedNumber, greaterThanOrEqualTo(16.0));	如果所测试的数值 testedNumber 大于或等于 16.0，则测试通过
hasItem	assertThat(iterableObject, hasItem("element"));	如果测试的迭代对象 iterableObject 含有元素 element 项，则测试通过
hasKey	assertThat(mapObject, hasKey("key"));	如果测试的 Map 对象 mapObject 含有键值 key，则测试通过
hasValue	assertThat(mapObject, hasValue("key"));	如果测试的 Map 对象 mapObject 含有元素值 value，则测试通过

下面通过示例讲解如何利用 assertThat 完成 Server 层的测试。

（1）创建 Spring Boot Web 应用项目 springboot0104，并添加相应的依赖。

（2）创建对应的类。

```
Student 类
@Component
public class Student {
    private String id;
    private String stuName;
    //省略 get 和 set 方法
}
StudentDB 类
    @Component
    public class StudentDB {
        private List<Student> lstStudent;
```

```java
    public List<Student> getAllStudents()
    {
        lstStudent = new ArrayList<>();
        Student s1 = new Student();
        s1.setId("20210101");
        s1.setStuName("张三");
        Student s2 = new Student();
        s2.setId("20210102");
        s2.setStuName("李四");
        lstStudent.add(s1);
        lstStudent.add(s2);
        return lstStudent;
    }
    public Student getStudentById(String id)
    {
        lstStudent = getAllStudents();
        for (Student student:lstStudent ) {
            if (student.getId().equals(id))
                return student;
        }
        return null;
    }
}
```

(3) 创建待测试的方法。

```java
@Service
public class StudentServiceImpl implements StudentService {
    @Autowired
    private StudentDB db;
    @Override
    public Student getStudentById(String id) {
        Student student = db.getStudentById(id);
        return student;
    }
}
```

(4) 引入类或 static 方法。

```java
import static org.junit.Assert.*;
import static org.hamcrest.Matchers.*;
```

(5) 编写测试方法。

```java
@RunWith(SpringRunner.class)
@SpringBootTest
public class StudentServiceImplTest {
  @Autowired
  private StudentService studentService;
  @Test
  public void getStudentById() {
        Student student = studentService.getStudentById("20210101");
//如果测试的结果相同则保持沉默,否则就抛出异常。
        Assert.assertSame(student.getStuName(), "张三");
        Assert.assertThat(student.getStuName(), is("张三 1"));
    }
}
```

(6) 测试结果。

选中单元测试方法 getStudentById()，右击 Run getStudentById()选项启动测试方法，此时控制台的打印消息如图 1-38 所示。

```
java.lang.AssertionError:
Expected: is "张三1"
     but: was "张三"
Expected :is "张三1"
     |
Actual   :"张三"
<Click to see difference>
```

图 1-38　getStudentById()单元测试方法效果

在实际项目编写单元测试的过程中，可以发现需要测试的类有很多依赖，这些依赖项又会有依赖，导致在单元测试代码里几乎无法完成构建，尤其是当依赖项尚未构建完成时会导致单元测试无法进行，如图 1-39 所示。

如果要对 A 进行测试，就必须对 B、C 测试，而对 B 测试就必须对其分支 D、E 进行测试，如果 D、E 没写完，那么 B 就会很难测试。要测试的目标类会有很多依赖，这些依赖的类、对象、资源又会有别的依赖，从而形成一个大的依赖树，要在单元测试的环境中完整地构建这样的依赖，是一件很困难的事情。

为此提出了 Mock 对象的概念，如图 1-40 所示。

图 1-39　测试原型　　　　　　图 1-40　Mock 图

将 B、C 所依赖的其他类和对象进行 Mock，构建它们的一个假对象，定义这些假对象上的行为，然后提供给被测试对象使用。这样无论 B、C 有多少分支依赖对象，都可以将其看成一个整体，而作为一个测试者只关心要测试的对象 A 所对应的分支对象就可以了。

下面通过示例讲解如何利用 Mockito 完成 Server 层的测试。

(1) 打开 Spring Boot 项目。

(2) Mockito 准备工作。

通过 Maven 进行管理，需要在项目的 pom.xml 文件中增加如下的依赖。

```xml
<dependency>
    <groupId>org.mockito</groupId>
    <artifactId>mockito-core</artifactId>
    <version>2.7.19</version>
    <scope>test</scope>
</dependency>
```

为了使代码更简洁，最好在测试类中导入静态资源。此外，为了使用常用的 junit 关键字，需要引入 JUnit 的两个类 Before 和 Test。

```java
import static org.mockito.Mockito.*;
import static org.junit.Assert.*;
import org.junit.Before;
import org.junit.Test;
```

(3) 创建对应的类和接口。

StudentDao 接口

```java
public interface StudentDao {
    Student getStudent(String id);
    boolean update(Student person);
```

```
}
StudentServiceImpl 类
@Service
public class StudentServiceImpl implements StudentService {
@Autowired
    private StudentDao studentDao;
    public boolean update(String id, String name) {
        Student student = studentDao.getStudent(id);
        if (student == null)
           { return false; }
        Student studentUpdate = new Student(student.getId(), name);
        return studentDao.update(student);
    }
}
```

（4）建立 StudentServiceImplTest 类。

```
import static org.junit.Assert.assertEquals;
import static org.mockito.ArgumentMatchers.isA;
import static org.mockito.Mockito.when;
import org.springframework.boot.test.mock.mockito.MockBean;
@RunWith(SpringRunner.class)
@SpringBootTest
public class Springboot0102ApplicationTests {
    @Autowired
    private StudentService studentService;
    @MockBean
    private StudentDao studentDao;
    @Before
    public void setUp() throws Exception {
```

设置对象调用的预期返回值，定义需要"假执行"的方法，并预先设定好该方法需要返回的值。当底层执行到该方法时，不会真正执行方法并返回自己预先设定好的值。需要注意的是，实际调用方法传入的参数必须跟预先设定"假执行"的方法参数一致，否则不生效。

```
        when(studentDao.getStudent("1")).thenReturn(new Student("1", "Person1"));
        when(studentDao.update(isA(Student.class))).thenReturn(true);
    }
    @Test
    public void testUpdate() throws Exception {
        boolean result = studentService.update("2", "new name");
        assertEquals(true, result);
    }
}
```

(5) 测试结果。

选中单元测试方法 testUpdate()，右击 Run testUpdate()选项启动测试方法，此时控制台的打印消息如图 1-41 所示。

```
java.lang.AssertionError:
Expected :true
Actual   :false
<Click to see difference>
```

图 1-41　testUpdate()单元测试方法效果

1.5.3　测试 Controller 层

在 Spring Boot 应用中，可以单独测试 Controller 代码，用来验证与 Controller 相关的

URL 路径映射、文件上传、参数绑定、参数校验等特性。可以通过@WebMvcTest 注解来完成 Controller 单元测试，当然也可以通过@SpringBootTest 测试 Controller。

MockMvc 是由 spring-test 包提供的，它实现了对 HTTP 请求的模拟，能够直接使用网络的形式转换到 Controller 的调用，使得测试速度变快且不依赖网络环境。同时，spring-test 包提供了一套验证的工具，对结果的验证十分方便。

MockMvc 的核心方法：

ResultActions MockMvc.perform(RequestBuilder requestBuilder)

RequestBuilder 用来构建 URL 请求，主要有以下几种请求。

- MockHttpServletRequestBuilder get(String urlTemplate, Object... urlVariables)：根据 URI 模板和 URI 变量值得到一个 GET 请求方式的 RequestBuilder，如果在 controller 的方法中选择 RequestMethod.GET 方式，那在 controllerTest 中对应就要使用 MockMvcRequestBuilders.get。
- post(String urlTemplate, Object... urlVariables)：与 GET 请求类似，但请求为 POST 方法。
- put(String urlTemplate, Object... urlVariables)：与 GET 请求类似，但请求为 PUT 方法。
- delete(String urlTemplate, Object... urlVariables)：与 GET 请求类似，但请求为 DELETE 方法。

对 ResultActions 有以下 3 种处理。

- ResultActions.andExpect：添加执行完成后的断言。添加 ResultMatcher 验证规则，验证控制器执行完成后结果是否正确。
- ResultActions.andDo：添加一个结果处理器，比如此处使用.andDo(MockMvcResultHandlers.print())输出整个响应结果信息，可以在调试的时候使用。
- ResultActions.andReturn：表示执行完成后返回相应的结果。

下面通过示例讲解如何利用 MockMvc 完成 Controller 层的测试。

(1) 打开 Spring Boot 项目。

(2) 创建 HelloWorldController 类。

```
@RestController
public class HelloWorldController {
    @Autowired
    private StudentService studentService;
    @GetMapping("/user/{userId}")
    @ResponseBody
    public Student getInfo(@PathVariable("userId") String userId){
        Student user = studentService.findByUserId(userId);
        return user;
    }
}
```

(3) 创建 StudentService 接口的实现类。

```
@Service
public class StudentServiceImpl implements StudentService {
    @Autowired
```

```
    private StudentDao studentDao;
    @Override
    public Student findByUserId(String userId) {
        Student student = studentDao.getStudent(userId);
        return student;
    }
}
```

（4）创建 HelloWorldControllerTest 测试类。

```
@RunWith(SpringRunner.class)
@WebMvcTest(HelloWorldController.class)
public class HelloWorldControllerTest {
@Autowired
private MockMvc mockMvc;
@MockBean
private StudentService studentService;
    @Test
    public void getInfo() throws Exception {
        Mockito.when(studentService.findByUserId("1")).thenReturn(new Student("1","张三"));
        MvcResult result = mockMvc.perform(get("/user/1")
        .contentType(MediaType.APPLICATION_JSON_UTF8)
        .accept(MediaType.APPLICATION_JSON_UTF8))
                .andExpect(status().isOk())
                .andReturn();
//获取数据
        JSONObject jsonObject = new JSONObject(result.getResponse().getContentAsString());
        Assert.assertThat(jsonObject.get("stuName"),is("张三1"));
    }
}
```

（5）测试结果。

选中单元测试方法 HelloWorldControllerTest()，右击 Run HelloWorldControllerTest()选项启动测试方法，此时控制台的打印消息如图 1-42 所示。

需要注意的是，在使用注解测试 Controller 时，带有@Service 和其他注解的组件类不会被自动扫描注册为 Spring 容器管理的 Bean，而@SpringBootTest 注解告诉 Spring Boot 去寻找一个主配置类（一个带@SpringBootApplication 的类），并使用它来启动 Spring 应用程序上下文，注入所有 Bean。另外，MockMvc 用来在 Servlet 容器内对 Controller 进行单元测试，并未真正发起 HTTP 请求调用 Controller。

```
java.lang.AssertionError:
Expected: is "张三1"
     but: was "张三"
Expected :is "张三1"

Actual   :"张三"
```

图 1-42　HelloWorldControllerTest()
　　　　单元测试方法效果

@WebMvcTest 用于从服务器端对 Controller 层进行统一测试。如果需要在客户端与应用程序交互时，应该使用@SpringBootTest 进行集成测试。

1.6　打包与部署

在通常情况下，部署单个项目会将项目打包成 WAR 包或 JAR 包，并将其部署到 Tomcat 或 Weblogic 等服务器上。如今 Docker 容器得到广泛应用，使得部署和发布变得更加快捷、自动化,且适应云平台环境。

Spring Boot 是一个快速开发框架,它简化了很多配置,甚至 Tomcat 服务器都内置在框架中。下面简单介绍 IDEA 如何简单快捷地打包项目,并部署到服务上。

Spring Boot 应用程序有以下两种运行方式。

- 以 JAR 包方式运行
- 以 WAR 包方式运行

1.6.1 以 JAR 包方式运行

1. 添加打包插件

在 pom.xml 文件中添加 Maven 打包插件,Spring Boot 使用该插件可以将项目打包成一个可运行的 JAR,而无须在目标服务器安装 Tomcat 等。

```
<build>
    <plugins>
        <plugin>
            <groupId>org.springframework.boot</groupId>
            <artifactId>spring-boot-maven-plugin</artifactId>
        </plugin>
    </plugins>
</build>
```

2. 使用 IDEA 开发工具进行打包

在图 1-43 所示的窗口中,单击右侧的 Maven Projects,打开对应的项目操作窗口,在弹出的窗口中选择 Lifecycle 目录,直接双击目录中的 package 选项,即可进行项目打包。

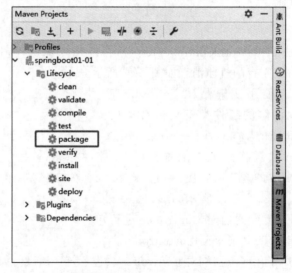

图 1-43 Maven Projects 窗口

根据上述操作说明,执行打包后,控制台就会显示打包运行过程及最终的打包结果,效果如图 1-44 所示。

在图 1-44 中,可以看到生成 JAR 包的存放路径和名称,以及编译成功的提示信息(BUILD SUCCESS)。

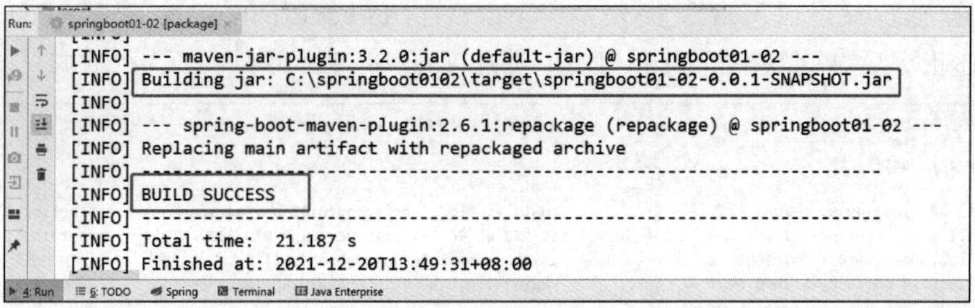

图 1-44 控制台输出结果

3. JAR 包位置

在工程的 target 目录下,可以看到 Maven 编译后的 JAR 包文件。也可以打开刚才打包后的所在位置,确实也有一个 JAR 包文件。

为了更加清楚 Spring Boot 中打包成的 JAR 包的具体目录结构,右击 JAR 包,使用解压缩软件打开并进入 BOOT-INF 目录中,如图 1-45 所示。

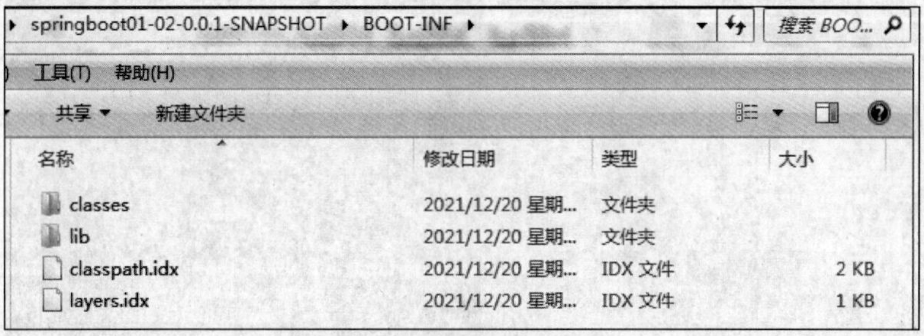

图 1-45 JAR 包目录结构

在图 1-45 中,BOOT-INF 目录下有 lib 和 classes 两个目录文件。其中,lib 目录下对应着所有添加的依赖文件导入的 JAR 文件;classes 目录下对应着项目打包编译后的所有文件。查看 BOOT-INF 目录下的 lib 目录,效果如图 1-46 所示。

图 1-46 lib 目录

在图 1-46 中,Spring Boot 项目打包成 JAR 包后,自动引入需要的各种 JAR 文件。lib 目录下还包括 Tomcat 相关的 JAR 文件,并且名字中包含 "tomcat-embed" 字样,这是 Spring Boot 内嵌的 JAR 包形式的 Tomcat 服务器。

4. JAR 包方式部署

用 IDEA 打开项目,在下方的工具控制台上切换到 Terminal 界面,该界面会默认打开项目所在的位置，然后输入指令 java -jar target\springboot01-02-0.0.1-SNAPSHOT.jar。
执行完上述指令后，就会启动 Spring Boot 项目，如图 1-47 所示。

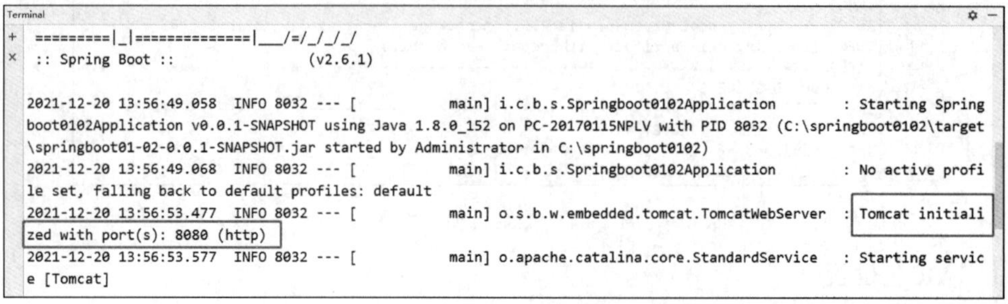

图 1-47　控制台输出结果

另外，还可以通过 cmd 启动命令窗口，进入项目生成的 JAR 包所在的路径，执行
java -jar springboot01-02-0.0.1-SNAPSHOT.jar 指令以运行项目，如图 1-48 所示。

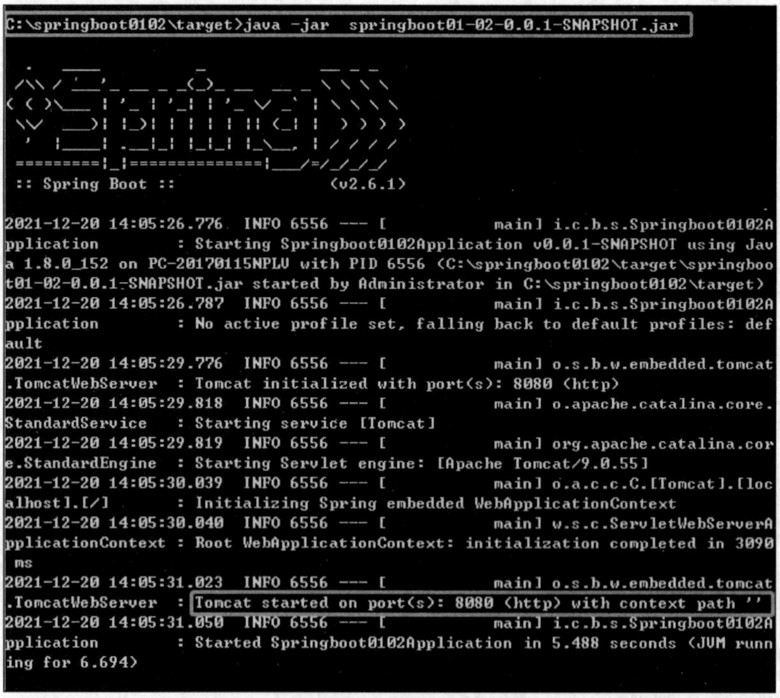

图 1-48　命令窗口下的 JAR 包部署

5. 访问项目

在浏览器地址栏中输入 http://localhost:8080/hello。

1.6.2　以 WAR 包方式运行

虽然通过 Spring Boot 内嵌的 Tomcat 可以直接将项目打包成 JAR 包进行部署，但有时还需要通过外部的可配置 Tomcat 进行项目管理，这就需要将项目打包成 WAR 包再进行部署。

1. 修改打包形式

在创建 Spring Boot 项目时,默认为 JAR 包,如果想要打包成 WAR 包,需要修改 pom.xml,将打包方式设置为 WAR。

```xml
<packaging>war</packaging>
```

2. 移除嵌入式 Tomcat 插件

在 Spring Boot 的 pom.xml 文件中,将 Web 启动器中的 Tomcat 依赖排除,这是因为打包的 WAR 项目要放在自己的 Tomcat 服务器中运行,需要排出 Spring Boot 项目内置的 Tomcat。然后手动将 Tomcat 插件加入依赖,并设置其 scope 值为 provided,如图 1-49 所示。

```xml
<dependency>
    <groupId>org.springframework.boot</groupId>
    <artifactId>spring-boot-starter-web</artifactId>
    <!-- 排除Web启动中自动依赖的Tomcat插件 -->
    <exclusions>
        <exclusion>
            <groupId>org.springframework.boot</groupId>
            <artifactId>spring-boot-starter-tomcat</artifactId>
        </exclusion>
    </exclusions>
</dependency>
<!--
手动依赖Tomcat插件,该依赖不会被打进去,目的主要是保证开发阶段本地Spring Boot
项目可以正常运行
-->
<dependency>
    <groupId>org.springframework.boot</groupId>
    <artifactId>spring-boot-starter-tomcat</artifactId>
    <!-- 打包时可以不用打包进去,其他设施会提供。该依赖理论上可以参与编译、测试、运行等周期,
    相当于compile,只是打包阶段做了exclude操作 -->
    <scope>provided</scope>
</dependency>
```

图 1-49 移除嵌入式 Tomcat 插件

3. 提供 Spring Boot 启动的 Servlet 初始化容器

提供 SpringBootServletInitializer 子类并覆盖其配置方法,这样做可以利用 Spring Framework 的 Servlet 3.0 支持,在 Servlet 容器启动应用程序时配置它。通常应该更新应用程序的主类以扩展 SpringBootServletInitializer,如图 1-50 所示。如果创建一个 SpringBootServletInitializer 子类的话,则这个子类应该与项目的启动类在同一级目录下。

```java
@SpringBootApplication
public class Springboot0102Application extends SpringBootServletInitializer {
    public static void main(String[] args) {
        SpringApplication.run(Springboot0102Application.class, args);
    }

    @Override
    protected SpringApplicationBuilder configure(SpringApplicationBuilder builder) {
        return builder.sources(Springboot0102Application.class);
    }
}
```

图 1-50 继承 SpringBootServletInitializer 的主类

4. 打包部署

WAR 包的打包方式与 JAR 包的打包方式一样。把 target 目录下的 WAR 包放到 Tomcat 的 webapps 目录下,启动 Tomcat 即可自动解压部署,如图 1-51 所示。

项目启动后访问路径会发生变化,如之前的访问路径是/hello,那么此时就需要加上 webapps 目录下解压的项目文件夹的名字,即/项目文件名/hello,如图 1-52 所示。

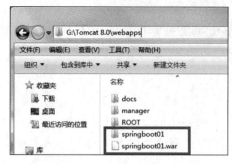

图 1-51　解压 WAR 包　　　　　　　图 1-52　运行项目

需要注意的是，如果需要打包为 WAR 包，那么不仅 POM 文件需要修改，应用程序也要做相应的改动。改动完成后，应用程序就无法在本地运行，需要将打包后的配置信息修改回来，这样不仅麻烦，而且容易出错。

在实际的项目中，并没有哪一种方式是最好的，要根据客户不同的需求制定不同的部署方案。例如，有些客户比较看中管理功能，要求数据源和 Tomcat 相关配置必须由管理员进行管理，则应选择 WAR 包方式；有些客户希望借助容器化进行大规模部署，则 JAR 方式更适合。

本章小结

本章主要对 Spring Boot 入门的基础知识进行讲解。首先，对 Spring Boot 概念和特点进行介绍，让读者快速了解 Spring Boot 框架的优势及学习的必要性；然后，通过使用 Maven、Spring Initializr 工具构建 Spring Boot 项目；最后，讲解了 Spring Boot 项目开发用到的单元测试和热部署。大家应该对 Spring Boot 整体的开发流程有一个初步认识，从而为后续学习 Spring Boot 做好铺垫。

在线测试

习题

一、单选题

1. 以下关于 Spring Boot 2.1.3 版本开发所需环境准备的说法，错误的是（　　）。
   A. 要求 Java 8 及以上版本的支持，同时兼容 Java 11
   B. Spring Boot 2.1.3 版本支持的第三方项目构建工具包括 Maven（3.3＋）和 Gradle（4.4＋）
   C. 目前 Java 项目支持的常用开发工具包括 Spring Tool Suite（STS）、Eclipse 和 IntelliJ IDEA 等
   D. Eclipse 是业界评价最高的一款 Java 开发工具，尤其在智能代码助手、重构、各类版本工具支持等方面

2. 以下关于 Spring Boot 概述的说法，错误的是（　　）。
   A. Pivotal 团队通过配置类的方式简化了 Spring 框架的使用，并开发了 Spring Boot 框架

B. Spring Boot 2.x 版本在 Spring Boot 1.x 版本的基础上进行了诸多功能的改进和扩展

C. Spring Boot 用于快速、敏捷地开发新一代基于 Spring 框架的应用，在开发过程中大量使用"约定优先配置"的思想

D. Spring Boot 并不是替代 Spring 框架的解决方案，而是与 Spring 框架紧密结合的用于提升 Spring 开发者体验的工具

3. 以下使用 Maven 方式创建 Spring Boot 项目的说法，正确的是(　　)。

A. 项目创建完成后，会默认打开创建 Maven 项目生成的 pom.xml 依赖文件

B. 项目创建完成后，会默认创建项目启动类

C. 创建项目后，打开 pom.xml 文件中的"Import Changes"会导入后续变化的依赖

D. 使用 Maven 方式创建 Spring Boot 项目需要手动添加依赖文件

4. 以下关于 Spring Boot 中单元测试的说法，正确的是(　　)。

A. Spring Boot 项目创建完成后，会自动生成单元测试类

B. Spring Boot 单元测试依赖为 test-spring-boot-starter

C. Spring Boot 单元测试类包括 @SpringBootTest 和 @RunWith 两个核心注解

D. Spring Boot 项目自动生成的测试类在 resources 目录下

二、填空题

1. Spring Boot 2.1.3 版本对 JDK 环境要求_____及以上版本的支持。

2. Spring Boot 项目中添加的测试依赖启动器，其<scope>范围默认为_____。

3. Spring Boot 2.1.3 版本官方声明支持的第三方项目构建工具包括_____和 Gradle(4.4+)。

4. 创建 Maven 项目过程中，GroupId 表示组织 ID，一般分为两个字段，包括_____和_____。

5. 注解_____用于标记 Spring Boot 测试类，并加载项目的 ApplicationContext 上下文环境。

三、简答题

Spring、Spring MVC、Spring Boot 三者有什么联系？为什么要学习 Spring Boot？

第2章 Spring Boot核心配置与注解

本章学习目标
- 熟悉 Spring Boot 自动化配置。
- 掌握 Spring Boot 配置文件属性值注入。
- 掌握 Spring Boot 多环境配置。

根据党的二十大报告,全面深化改革要坚持从大局出发考虑问题,并具有全局性、长远性、系统性、前瞻性的战略思维。

本章通过学习微服务项目架构组件,构建 Spring Boot 项目的配置文件、配置类、多环境下的部署方式,并进而深刻理解全局考虑的战略思维。

2.1 自动化配置

在常规的 Spring 应用程序中,充斥着大量的配置文件,需要手动配置这些文件,如配置组件扫描、配置 Servlet、配置视图解析器和配置 HTTP 编码等。以配置 HTTP 编码为例,需要在 web.xml 文件中配置类似的一个 Filter:

```xml
<filter>
    <filter-name>CharacterEncodingFilter</filter-name>
    <filter-class>org.springframework.web.filter.CharacterEncodingFilter</filter-class>
    <init-param>
        <param-name>encoding</param-name>
        <param-value>UTF-8</param-value>
    </init-param>
</filter>
```

上述的配置可以称得上是 Spring 应用程序的经典案例,常规的配置让开发人员将更多的经历耗费在配置文件上。这些配置都是一些固定模式的配置方式,甚至很多都是模板代码。既然如此,有没有一种可能,让 Spring 自动完成这些模板配置工作呢?答案是肯定的,这就是 Spring Boot 自动化配置产生的初衷。将开发人员从繁重的配置工作中解放出来,而这些烦琐的配置细节交由 Spring Boot 去完成,如果需要提供自己的配置参数,只需要覆盖自动配置的参数即可。

在开发任何一个 Spring Boot 项目时,都会用到如下的启动类:

@SpringBootApplication

```
public class Application {
    public static void main(String[] args) {
        SpringApplication.run(Application.class, args);
    }
}
```

从上面的代码可以看出，Annotation 定义（@SpringBootApplication）和类定义（SpringApplication.run）最为关键，所以要揭开 Spring Boot 的神秘面纱，就要从这两位开始。

2.1.1 @SpringBootApplication

在项目的启动类上需要添加注解@SpringBootApplication，这是 Spring Boot 的核心注解，目的是开启自动化配置，它其实是一个组合注解。

```
@Target(ElementType.TYPE)
@Retention(RetentionPolicy.RUNTIME)
@Documented
@Inherited
@SpringBootConfiguration
@EnableAutoConfiguration
@ComponentScan(excludeFilters = {
    @Filter(type = FilterType.CUSTOM, classes = TypeExcludeFilter.class),
    @Filter(type = FilterType.CUSTOM, classes = AutoConfigurationExcludeFilter.class) })
public @interface SpringBootApplication {
    ...
}
```

虽然定义使用了多个 Annotation 进行原信息标注，但实际上重要的只有 3 个 Annotation。

- @Configuration（@SpringBootConfiguration 里面还是应用了@Configuration）
- @EnableAutoConfiguration
- @ComponentScan

1. @Configuration

这里的 @Configuration 就是 JavaConfig 形式的 Spring IoC 容器的配置类使用的 @Configuration，Spring Boot 社区推荐使用基于 JavaConfig 的配置形式，所以，这里的启动类标注了@Configuration 后，本身其实也是一个 IoC 容器的配置类。

举个简单例子回顾 XML 跟 Config 配置方式的区别。

基于 XML 的配置形式：

```
<bean id = "studentService" class = "it.server.StudentServiceImpl">
    ...
</bean>
```

而基于 JavaConfig 的配置形式：

```
@Configuration
public class StudentConfiguration{
    @Bean
    public StudentService studentService(){
        return new StudentServiceImpl();
```

    }
}
```

在Spring Boot中，通过使用@Configuration注解指定自定义配置类。配置类中@Bean的方法，其返回值将作为一个Bean定义注册到Spring的IoC容器，方法名默认为该Bean定义的id。

#### 2. @ComponentScan

@ComponentScan这个注解在Spring中很重要，它对应XML配置中的元素。@ComponentScan的功能其实就是自动扫描并加载符合条件的组件（如@Component和@Repository等）或者Bean定义，将这些Bean定义加载到IoC容器中，其作用等同于<context:component-scan base-package="it.com.boot.springboot0104"/>配置。

@ComponentScan的常用属性如下：

- basePackages、value：指定扫描路径，如果为空则以@ComponentScan注解类所在的包为基本的扫描路径。
- basePackageClasses：指定具体扫描的类。
- includeFilters：指定满足Filter条件的类。
- excludeFilters：指定排除Filter条件的类。

可以通过basePackages等属性来精确地定制@ComponentScan自动扫描的范围，如果不指定，则默认Spring框架实现会从声明@ComponentScan所在类的package进行扫描。Spring Boot的启动类最好是放在root package下，因为默认不指定basePackages。

#### 3. @EnableAutoConfiguration

@EnableAutoConfiguration注解表示开启自动配置功能，该注解是Spring Boot框架最重要的注解，也是实现自动化配置注解。@EnableAutoConfiguration会根据类路径中的JAR依赖为项目进行自动配置，如添加spring-boot-starter-web依赖后，会自动添加Tomcat和Spring MVC的依赖，Spring Boot会对Tomcat和Spring MVC进行自动配置。

@EnableAutoConfiguration的核心代码如下。

```
@Target(ElementType.TYPE)
@Retention(RetentionPolicy.RUNTIME)
@Documented
@Inherited
@AutoConfigurationPackage
@Import(EnableAutoConfigurationImportSelector.class)
public @interface EnableAutoConfiguration {
}
```

其中最重要的是@Import注解，@EnableAutoConfiguration就是借助@Import的帮助，将所有符合自动配置条件的Bean定义加载到IoC容器中。而EnableAutoConfigurationImportSelector内部则使用了SpringFactoriesLoader.loadFactoryNames方法，扫描具有META-INF/spring.factories文件的JAR包。

因此自动化配置就像一只"八爪鱼"一样，如图2-1所示。借助于Spring框架原有工具类SpringFactoriesLoader的支持，@EnableAutoConfiguration可以实现智能地自动配置。
```

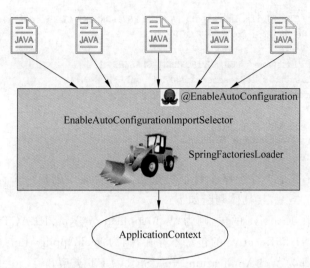

图 2-1　EnableAutoConfiguration 得以生效的关键组件关系图

2.1.2　SpringApplication

每个 Spring Boot 项目都有一个主程序启动类，在主程序启动类中有一个启动项目的 main() 方法，在该方法中通过执行 SpringApplication.run() 即可启动整个 Spring Boot 程序。那么 SpringApplication.run() 方法是如何启动 Spring Boot 项目的呢？

1. SpringApplication 初始化阶段

进入 SpringApplication.run 方法，创建一个 SpringApplication 对象，其构造函数如下：

```
public SpringApplication(ResourceLoader resourceLoader, Class<?>...primarySources) {
    this.resourceLoader = resourceLoader;
    Assert.notNull(primarySources, "PrimarySources must not be null");
    //primarySources 为 run 方法传入的引导类
    this.primarySources = new LinkedHashSet<>(Arrays.asList(primarySources));
    //推断 Web 应用类
    this.webApplicationType = deduceWebApplicationType();
    //初始化 initializers 属性
    setInitializers((Collection) getSpringFactoriesInstances(
            ApplicationContextInitializer.class));
    //初始化监听器
    setListeners((Collection) getSpringFactoriesInstances(ApplicationListener.class));
    //推断应用引导类
    this.mainApplicationClass = deduceMainApplicationClass();
}
```

推断 Web 应用类的方法 deduceWebApplicationType() 如下：

```
private static final String[] WEB_ENVIRONMENT_CLASSES = { "javax.servlet.Servlet","org.springframework.web.context.ConfigurableWebApplicationContext" };
private static final String REACTIVE_WEB_ENVIRONMENT_CLASS = "org.springframework.web.reactive.DispatcherHandler";
private static final String MVC_WEB_ENVIRONMENT_CLASS = "org.springframework.web.servlet.DispatcherServlet";
private WebApplicationType deduceWebApplicationType() {
    if (ClassUtils.isPresent(REACTIVE_WEB_ENVIRONMENT_CLASS, null)
```

```java
                && !ClassUtils.isPresent(MVC_WEB_ENVIRONMENT_CLASS, null)) {
            return WebApplicationType.REACTIVE;
        }
        for (String className : WEB_ENVIRONMENT_CLASSES) {
            if (!ClassUtils.isPresent(className, null)) {
                return WebApplicationType.NONE;
            }
        }
        return WebApplicationType.SERVLET;
    }
```

根据 classpath 下是否存在某个特征类来决定是否应该创建一个为 Web 应用使用的 ApplicationContext 类型,其具体判断如下:

(1) 如果仅存在 Reactive 的包,则为 WebApplicationType.REACTIVE 类型。

(2) 如果 Servlet 和 Reactive 的包都不存在,则为 WebApplicationType.NONE 类型。

(3) 其他情况都为 WebApplicationType.SERVLET 类型。

接下来分析看初始化 initializers 属性的过程,其通过 getSpringFactoriesInstances (ApplicationContextInitializer.class)方法获取初始化器。

```java
private <T> Collection<T> getSpringFactoriesInstances(Class<T> type,
        Class<?>[] parameterTypes, Object... args) {
    ClassLoader classLoader = Thread.currentThread().getContextClassLoader();
    //Use names and ensure unique to protect against duplicates
    Set<String> names = new LinkedHashSet<>(
            SpringFactoriesLoader.loadFactoryNames(type, classLoader));
    List<T> instances = createSpringFactoriesInstances(type, parameterTypes,
            classLoader, args, names);
    AnnotationAwareOrderComparator.sort(instances);
    return instances;
}
```

该方法流程如下:

(1) 通过 SpringFactoriesLoader.loadFactoryNames(type,classLoader)方法,在 META-INF/spring.factories 文件下查找 ApplicationContextInitializer 类型对应的资源名称。

(2) 实例化上面的资源信息(初始化器)。

(3) 对初始化器根据 Ordered 接口或者@Order 注解进行排序。

同理,初始化 listeners 监听器也是类似的流程。

SpringApplication 初始化阶段的最后一步是推断引导类 deduceMainApplicationClass():

```java
private Class<?> deduceMainApplicationClass() {
    try {
        StackTraceElement[] stackTrace = new RuntimeException().getStackTrace();
        for (StackTraceElement stackTraceElement : stackTrace) {
            if ("main".equals(stackTraceElement.getMethodName())) {
                return Class.forName(stackTraceElement.getClassName());
            }
        }
    }
    catch (ClassNotFoundException ex) {
        //Swallow and continue
```

```
        }
        return null;
}
```

该阶段将调用栈中 main 方法所在的类作为引导类。

2. SpringApplication 运行阶段

SpringApplication 运行阶段属于核心过程，完全围绕 run(String…)方法展开。该过程结合初始化阶段完成的状态，进一步完善运行时所需要准备的资源，随后启动 Spring 应用上下文。在此期间伴随着 Spring Boot 和 Spring 事件的触发，形成完整的 SpringApplication 生命周期。因此，下面将围绕以下 3 个子议题进行讨论。

- SpringApplication 准备阶段
- ApplicationContext 启动阶段
- ApplicationContext 启动后阶段

```
public ConfigurableApplicationContext run(String... args) {
    StopWatch stopWatch = new StopWatch();
    stopWatch.start();
    ConfigurableApplicationContext context = null;
    FailureAnalyzers analyzers = null;
    //配置属性
    configureHeadlessProperty();
    //获取监听器
    //利用 loadFactoryNames 方法从路径 MEAT-INF/spring.factories 中找到所有的
    SpringApplicationRunListener
    SpringApplicationRunListeners listeners = getRunListeners(args);
    //启动监听
    //调用每个 SpringApplicationRunListener 的 starting 方法
    listeners.starting();
    try {
        //将参数封装到 ApplicationArguments 对象中
        ApplicationArguments applicationArguments = new DefaultApplicationArguments( args);
        //准备环境
        //触发监听事件——调用每个 SpringApplicationRunListener 的 environmentPrepared 方法
        ConfigurableEnvironment environment = prepareEnvironment(listeners,applicationArguments);
        //从环境中取出 Banner 并打印
        Banner printedBanner = printBanner(environment);
        //依据是否为 Web 环境创建 Web 容器或者普通的 IoC 容器
        context = createApplicationContext();
        analyzers = new FailureAnalyzers(context);
        //准备上下文
        //(1)将 environment 保存到容器中
        //(2)触发监听事件——调用每个 SpringApplicationRunListeners 的 contextPrepared 方法
        //(3)调用 ConfigurableListableBeanFactory 的 registerSingleton 方法向容器中注入
applicationArguments 与 printedBanner
        //(4)触发监听事件——调用每个 SpringApplicationRunListeners 的 contextLoaded 方法
        prepareContext(context, environment, listeners, applicationArguments,printedBanner);
        //刷新容器，完成组件的扫描、创建、加载等
        refreshContext(context);
        afterRefresh(context, applicationArguments);
        //触发监听事件——调用每个 SpringApplicationRunListener 的 finished 方法
        listeners.finished(context, null);
        stopWatch.stop();
```

```
            if (this.logStartupInfo) {
                new StartupInfoLogger(this.mainApplicationClass).logStarted(getApplicationLog(),
stopWatch);
            }
            //返回容器
            return context;
        }
        catch (Throwable ex) {
            handleRunFailure(context, listeners, analyzers, ex);
            throw new IllegalStateException(ex);
        }
    }
```

Spring Boot 应用启动的执行流程如图 2-2 所示，具体步骤如下。

（1）初始化 SpringApplication 实例：确定 Web 应用类型、加载初始化器和监听器、推断 main 方法的定义类。

（2）通过 SpringFactoriesLoader 加载的 SpringApplicationRunListener，调用它们的 started 方法。

（3）创建并配置当前 Spring Boot 应用将要使用的 Environment，如 application.properties 文件和外部配置。

（4）根据 Web 服务类型创建不同的 Spring 应用上下文，并将之前准备好的 Environment 设置给 Spring 应用上下文 ApplicationContext 使用。

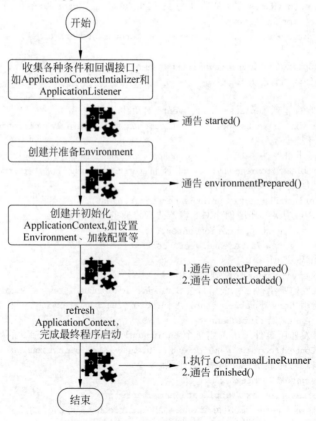

图 2-2　Spring Boot 应用启动的执行流程

(5) 遍历初始化器，对 ApplicationContext 进行初始化操作。

(6) 初始化上下文 data handling subsystem()，在 invokeBeanFactoryPostProcessors 中加载所有资源，如 Configuration Class、类名、包名和 Spring XML 配置资源路径。将所有 BeanDefinition 加载至 ApplicationContext，并进行自动装配，完成初始化 IoC 容器等操作。

(7) 寻找当前 ApplicationContext 中是否注册有 CommandLineRunner 或者 ApplicationRunner，如果有则遍历执行它们。

2.2 全局配置

Spring Boot 避免了大部分手动配置，但是对于一些特殊的情况，还是需要进行手动配置的。Spring Boot 项目提供了一个全局配置文件 application.properties 或 application.yml，对一些默认的配置值进行修改，存放在 src/main/resources 目录下或类路径的 /config 目录下。

application.properties 配置文件比较简单，其形式如下：

```
key = value
```

在创建 Spring Boot 工程时，内置的 Tomcat 默认端口为 8080。有时需要重新制定端口，使用 server.port 就可以制定内置 Tomcat 容器的端口：

```
server.port = 8088
```

如果开发者想设置一个 Web 应用程序的上下文路径，可以在 application.properties 文件中配置如下内容：

```
server.servlet.context-path = /boot
```

这时应该在浏览器地址栏中输入 http://localhost:8088/boot/hello，访问如下控制器类中的请求处理方法：

```
@RestController
public class HelloController {
    @GetMapping("/hello")
    public String hello() {
        return "欢迎来到 SpringBoot 世界";
    }
}
```

YAML 文件格式是 Spring Boot 支持的一种 JSON 超集文件格式，相较于传统的 Properties 配置文件，YAML 文件以数据为核心，是一种更为直观且容易被计算机识别的数据序列化格式。application.yml 配置文件的工作原理和 application.properties 是一样的，只不过 YAML 格式配置文件看起来更简洁一些。YAML 文件的扩展名可以使用 .yml 或者 .yaml。

application.yml 文件使用"key:（空格）value"格式配置属性，以空格的缩进来控制层级关系，即左对齐的一列数据为同一个层级。

```
server:
  port: 8888
  path: /path
```

```
name:
  first: tian
  second: song
```

YAML 支持的数据结构有以下 3 种。

- 字面量：单个的、不可再分的值。
- 对象：键值对的集合，又称为映射（mapping）/哈希（hashes）/字典（dictionary of Chinese characters）。
- 数组：一组按次序排列的值，又称为序列（sequence）/列表（tabulation）。

1. 字面量

当 YAML 配置文件中配置的属性值为普通数据类型时，可以直接配置对应的属性值。同时，对于字符串类型的属性值，默认不用加单引号或双引号。

在双引号里一旦加入特殊字符，则不会转义字符串里面的特殊字符，特殊字符会表示原本的含义。

```
name: "zhangsan \n lisi"
```

输出：

```
zhangsan
lisi
```

在单引号里加入特殊字符时，特殊字符最终只是一个普通的字符串数据。

```
name: 'zhangsan \n lisi'
```

输出：

```
zhangs\nlisi
```

2. YAML 对象

对象键值对使用冒号结构表示，即 key:（空格）value，冒号后面要加一个空格。也可以使用 key:{key1:value1, key2:value2,…} 表示。

使用缩进表示层级关系为

```
key:
    child-key: value
    child-key2: value2
```

配置类中的字段为

```
private Map<String,String> maps;
private Birthday birthday;
```

在 YML 配置文件中，行内写法为

```
person.maps: {math: 80, english: 90}
person.birthday: {year: 2000,month: 8 ,day :1}
```

其中，需要注意冒号后的空格。

另一种行内写法为

```
person:
  maps:
    math: 80
```

```
          english:90
       birthday:
          year: 2000
          month:8
          day:1
```

3. 数组

以"-"开头的行表示构成一个数组。

```
- Cat
- Dog
- Goldfish
```

数组也可以采用行内表示法。

```
animal: [Cat, Dog,Goldfish]
```

在 Spring Boot 中，配置文件可以是 properties、yaml、yml 这 3 个格式中的任意一个，其中 properties 是键值对形式的，yaml 和 yml 其实是同一种格式，只是后缀名不同而已。当 3 种文件同时存在时，其实 3 个文件中的配置信息都会生效。但是当 3 个文件中有配置信息冲突时，逻辑顺序是依次加载 yml、yaml、properties，也就是 properties 里配置的内容会覆盖另外两个的配置。至于原因可以在 pom.xml 文件中找到，进入 spring-boot-starter-parent 依赖内部，如图 2-3 所示。

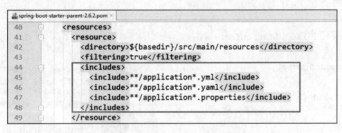

图 2-3　配置文件加载顺序

2.3　自定义配置

Spring Boot 提供了 application.properties 配置文件，可以进行自定义配置或对默认的配置进行修改，以适应具体的生产情况。当然，Spring Boot 还包括一些第三方的配置，接下来介绍如何读取第三方的配置信息。

2.3.1　注入自定义属性到字段中

@Value 注解用于读取 Java 代码中的环境或应用程序的属性值，通常用来配置单个环境变量。@Value 注解读取属性值的语法如下：

```
@Value(" ${property_key_name}")
```

接下来通过示例讲解@Value 注解的使用方法。

（1）新建一个名为 springboot0201 的项目，选择 Web 依赖。

（2）在 application.properties 配置文件中添加自定义属性 student.name，代码如下：

视频讲解

student.name=张三

（3）在it.com.boot.springboot0201.po包下新建一个Student实体类，并使用@Value注解注入属性。

```
@Component
public class Student {
    @Value("${student.name}")
    private String stuName;
    @Value("24")
    private Integer stuAge;
    @Value("${student.sex:男}")
    private String stuSex;
    //省略toString方法
}
```

@Value不仅支持注入属性值，而且支持直接为属性赋值。

注意：如果在运行应用程序时未找到对应的属性，则Spring Boot会抛出"非法参数"异常（因为无法解析值"${student.sex}"中的占位符"student.sex"）。

要解决占位符问题，可以使用下面给出的代码，为属性设置默认值。

```
@Value("${student.sex:男}")
```

（4）编写测试文件。

```
@SpringBootTest
class Springboot0201ApplicationTests {
    @Autowired
    private Student student;
    @Test
    void contextLoads() {
        System.out.println(student.toString());
    }
}
```

```
2022-01-07 09:59:58.397  INFO 7936 --- [
2022-01-07 09:59:58.399  INFO 7936 --- [
2022-01-07 10:00:07.258  INFO 7936 --- [
Student{stuName='张三', stuAge=24, stuSex='男'}
```

图2-4　contextLoads()方法执行结果

（5）运行contextLoads()方法，控制台的输出结果如图2-4所示。

2.3.2　注入自定义属性到对象中

视频讲解

在Spring Boot中，当需要获取配置文件数据时，除了可以用Spring自带的@Value注解外，Spring Boot还提供了一种更加方便的方式：@ConfigurationProperties。只要在Bean上添加这个注解，指定好配置文件的前缀，那么对应的配置文件数据就会自动填充到Bean中。

接下来通过示例讲解@ConfigurationProperties注解的使用方法。

（1）新建一个名为springboot0202的项目，选择Web依赖。

（2）在it.com.boot.springboot0202.po包下新建Student和Birthday实体类，并使用@ConfigurationProperties(prefix="student")注解注入属性。

```
@Component
public class Birthday {
    private Integer year,month,day;
    //省略set和get方法
    //省略toString方法
```

```
}
@Component
@ConfigurationProperties(prefix = "student")
public class Student {
    private String stuName;
    private Integer stuAge;
    private String stuSex;
    private Map<String, Integer> score;
    private List<String> address;
    private Birthday birthday;
    //省略 set 和 get 方法
    //省略 toString 方法
}
```

prefix="student"：声明配置前缀，主要作用是将该前缀下的所有属性进行映射。

(3) 在 application.yaml 配置文件中添加自定义属性，代码如下。

```
student:
 stu-name: 张三
 stu-age: 26
 stu-sex: 男
 address: [天津,北京,上海]
 score: {math: 90,english: 80,computer: 95}
 birthday: {year: 1999,month: 1,day: 2}
```

(4) 配置自动提示。

在配置自定义属性时，如果想要获得和配置 Spring Boot 属性自动提示一样的功能，则需要在 pom.xml 文件中加入下面的依赖。

```
<!-- 导入配置文件处理器,配置文件进行绑定就会有提示 -->
<dependency>
    <groupId>org.springframework.boot</groupId>
    <artifactId>spring-boot-configuration-processor</artifactId>
    <optional>true</optional>
</dependency>
```

(5) 编写 JSON 文件。

配置 target/classes/META-INF/spring-configuration-metadata.json 文件来描述会产生大量代码，为此可以通过 IDEA 来生成代码。

在 IDEA 中依次单击 file→Settings 选项，然后搜索 Annotation Processors，最后勾选 Enable annonation processing 就完成了。

使用 Maven 工具依次执行 clean 和 compile 命令，即可在编译后的文件中看到自动生成的 spring-configuration-metadata.json，部分代码如下。

```
"groups": [
    {
        "name": "student",
        "type": "it.com.boot.springboot0202.po.Student",
        "sourceType": "it.com.boot.springboot0202.po.Student"
    }
],
"properties": [
```

```
        {
            "name": "student.address",
            "type": "java.util.List<java.lang.String>",
            "sourceType": "it.com.boot.springboot0202.po.Student"
        },
//省略其他属性
]
```

（6）编写测试文件。

```
@SpringBootTest
class Springboot0202ApplicationTests {
    @Autowired
    private Student student;
    @Test
    void contextLoads() {
        System.out.println(student.toString());
    }
}
```

运行contextLoads()方法，控制台的输出结果如图2-5所示。

```
2022-01-07 11:18:49.307  INFO 8432 --- [           main] i.c.b.s.Springboot0204ApplicationTests   : Starting Springboot0204ApplicationTests using Java 1.8.0_152 on PC-
2022-01-07 11:18:49.310  INFO 8432 --- [           main] i.c.b.s.Springboot0204ApplicationTests   : No active profile set, falling back to default profiles: default
2022-01-07 11:18:55.478  INFO 8432 --- [           main] i.c.b.s.Springboot0204ApplicationTests   : Started Springboot0204ApplicationTests in 7.176 seconds (JVM runnin
Student{stuName='张三', stuAge=26, stuSex='男', score={math=90, english=80, computer=95}, address=[天津, 北京, 上海], birthday=Birthday{year=1999, month=1, day=2}}
```

图2-5　contextLoads()方法执行结果

@ConfigurationProperties和@value有着相同的功能。如果只是在某个业务逻辑中需要获取配置文件中的某项值，则使用@Value；如果专门编写了一个JavaBean来与配置文件进行映射，则直接使用@ConfigurationProperties。

@ConfigurationProperties的POJO类的命名比较严格，因为它必须和prefix的后缀名保持一致，否则无法绑定值。Spring Boot会使用一些宽松的规则将环境属性绑定到@ConfigurationProperties所注解类的属性上，因此环境属性名和类中的属性名之间不需要完全匹配。例如，特殊的后缀名是"stu-name"这种带"半字线"的情况，在POJO里面的命名规则是下画线转驼峰模式就可以绑定成功，即"stuName"。常见的松散绑定规则如表2-1所示。

表2-1　常见的松散绑定规则

属性文件中配置	说　　明
student.stu-name	羊肉串模式case，推荐使用
student.stuName	标准驼峰模式
student.stu_name	下画线模式
STUDENT.STU_NAME	大写下画线，使用系统环境时推荐使用

2.3.3　注入自定义配置文件

视频讲解

所有的配置都可以写入application.properties文件中，该文件会被Spring Boot自动加载。但实际上，很多时候需要自定义配置文件，这些文件就需要进行手动加载，Spring Boot是不会自动识别这些文件的。此时可以使用@PropertySource注解找到项目的其他配置文件，然后结合2.3.1节和2.3.2节中任意一种方法读取即可。

接下来通过示例讲解@PropertySource注解的使用方法。

(1) 新建一个名为 springboot0203 的项目,选择 Web 依赖。

(2) 在 it.com.boot.springboot0203.po 包下新建 Student 和 Birthday 实体类,并使用 @ConfigurationProperties(prefix = "student") 注解注入属性,具体见 2.3.2 节的代码。

(3) 在 resources 文件夹下创建 student.properties 自定义配置文件,代码如下。

```
student.stuName = 张三
student.stuAge = 26
student.stuSex = 男
student.address = 天津,北京,上海
student.score.math = 90
student.score.english = 80
student.score.computer = 95
student.birthday.year = 1999
student.birthday.month = 1
student.birthday.day = 2
```

(4) 修改 Student 类,加入 @PropertySource 注解指定自定义配置文件位置和名称,代码如下。

```
@Component
@ConfigurationProperties(prefix = "student")
@PropertySource("classpath:student.properties")
public class Student {
    //省略 Student 类
}
```

(5) 使用 @ConfigurationProperties 注解进行配置文件属性值注入时,支持 JSR3.0 数据校验,对注入的值进行一些简单的校验,示例代码如下。

```
<!-- 在 pom.xml 中添加 JSR3.0 数据校验依赖 -->
<dependency>
    <groupId>org.springframework.boot</groupId>
    <artifactId>spring-boot-starter-validation</artifactId>
</dependency>
@Component
@ConfigurationProperties(prefix = "student")
@PropertySource("classpath:student.properties")
@Validated
public class Student {
    @Range(min = 20, max = 30, message = "年龄必须在 20~30 岁")
    private Integer stuAge;
    //其他属性见源代码
}
```

@Range 注解会对 stuAge 字段的注入值进行检验,如果注入年龄不在 20~30 岁,则不是一个合法的年龄,就会抛出异常。

JSR 3.0 常用数据校验如表 2-2 所示。

表 2-2 JSR 3.0 常用数据校验

注　　解	注　解　详　情
@Null	被指定的注解元素必须为 Null
@NotNull	任意类型,不能为 Null,但可以为空,如空数组、空字符串

续表

注 解	注 解 详 情
@NotBlank	针对字符串，不能为 Null，且去除前后空格后的字符串长度要大于 0
@NotEmpty	针对字符串、集合、数组，不能为 Null，且长度要大于 0
@Size	针对字符串、集合、数组，判断长度是否在给定范围内
@Length	针对字符串，判断长度是否在给定范围内
@AssertTrue	针对布尔值，用来判断布尔值是否为 true
@AssertFalse	针对布尔值，用来判断布尔值是否为 false
@Past	针对日期，用来判断当前日期是否为过去的日期
@Future	针对日期，用来判断当前日期是否为未来的日期
@Max(value)	针对字符串、数值，用来判断是否小于或等于某个指定值
@Min(value)	针对字符串、数值，用来判断是否大于或等于某个指定值
@Range(min=，max=)	被指定的元素必须在合适的范围内
@Pattern	验证字符串是否满足正则表达式
@Email	验证字符串是否满足邮件格式
@Url	验证是否满足 URL 格式
@Validate	对实体类进行校验

（6）编写并运行测试文件。

@PropertySource 可用于加载指定的配置文件，可以是 properties 配置文件，也可以是 yml、yaml 配置文件。但在加载 yml 配置文件时，会出现无法获取属性的问题。修改 springboot0203 项目中的配置文件为 student.yaml，再次运行测试程序，如图 2-6 所示。

```
2022-01-07 10:22:06.031  INFO 8180 --- [           main] i.c.b.s.Springboot0204ApplicationTest
2022-01-07 10:22:06.035  INFO 8180 --- [           main] i.c.b.s.Springboot0204ApplicationTest
2022-01-07 10:22:14.534  INFO 8180 --- [           main] i.c.b.s.Springboot0204ApplicationTest
Student{stuName='null', stuAge=null, stuSex='null', score=null, address=null, birthday=null}
```

图 2-6 加载 yml 配置文件失败

所有属性值都为空，显然是没有读取到配置文件，这说明@PropertySource 默认不支持 yml、yaml 读取。通过查看@PropertySource 源码可以看出，@PropertySource 的注解中有一个 factory 属性，可指定一个自定义的 PropertySourceFactory 接口实现，用于解析指定的文件。默认的实现是 DefaultPropertySourceFactory，且使用 PropertiesLoaderUtils.loadProperties 进行文件解析，这个方法是一行一行地读取，然后根据冒号、等于号、空格等进行校验，经过一系列遍历后获取 key 和 value，而 yaml 语法是以缩进来辨别的，所以默认使用 Properties 进行解析。因此，要让@PropertySource 支持 yml、yaml 读取的话，就需要重新模仿 DefaultPropertySourceFactory 写一个 yaml 资源文件读取的工厂类。

接下来通过示例讲解@PropertySource 注解如何支持 yaml 资源文件的读写。

（1）新建一个名为 springboot0204 的项目，选择 Web 依赖。

（2）在 resources 文件夹下创建 student.yaml 自定义配置文件，代码如下。

```
student:
  stu-name: 张三
  stu-age: 26
  stu-sex: 男
  address: [天津,北京,上海]
  score: {math: 90,english: 80,computer: 95}
  birthday: {year: 1999,month: 1,day: 2}
```

（3）新建 it.com.boot.springboot0204.utils 包，在该包下新建 YamlPropertyResourceFactory 类并继承 PropertySourceFactory 接口，代码如下。

```
public class YamlPropertyResourceFactory implements PropertySourceFactory {
    @Override
    public PropertySource<?> createPropertySource(@Nullable String name, EncodedResource encodedResource) throws IOException {
        String resourceName = Optional.ofNullable(name).orElse(encodedResource.getResource().getFilename());
        if (resourceName.endsWith(".yml") || resourceName.endsWith(".yaml")) {//yaml 资源文件
            List<PropertySource<?>> yamlSources = new YamlPropertySourceLoader().load(resourceName, encodedResource.getResource());
            return yamlSources.get(0);
        } else {
            return new PropertiesPropertySource(resourceName, new Properties());
        }
    }
}
```

（4）在 it.com.boot.springboot0204.po 包下新建 Student 和 Birthday 实体类，并使用 @ConfigurationProperties 和 @PropertySource 注解，代码如下。

```
@Component
@ConfigurationProperties(prefix = "student")
@PropertySource(value = "classpath:student.yaml", encoding = "utf-8", factory = YamlPropertyResourceFactory.class)
@Validated
public class Student {
    //其他属性和方法见源代码
}
```

（5）编写测试类并运行，运行结果如图 2-5 所示。

2.3.4　自动扫描配置类

在 Spring Boot 框架中，推荐使用配置类的方式向容器中添加和配置组件。在 Spring Boot 框架中，通常使用 @Configuration 注解定义一个配置类，Spring Boot 会自动扫描和识别配置类，从而替换传统 Spring 框架中的 XML 配置文件。

当定义一个配置类后，还需要在类中的方法上使用 @Bean 注解进行组件配置，将方法的返回对象注入 Spring 容器中，并且组件名称默认使用的是方法名，当然也可以使用 @Bean 注解的 name 或 value 属性自定义组件的名称。

接下来通过示例讲解 @Configuration 编写自定义配置类的用法。

（1）新建一个名为 springboot0205 的项目，选择 Web 依赖。

（2）在 resources 文件夹下创建 application.properties 自定义配置文件，代码参见 2.3.3 节案例。

（3）在 it.com.boot.springboot0205.po 包下新建 Student 和 Birthday 实体类，此时两个类不需要加入任何注解，代码参见 2.3.3 节案例。

（4）新建 it.com.boot.springboot0205.config 包，在该包下新建 AppConfig 类，代码如下。

```
@Configuration
public class AppConfig {
    @Bean
    @ConfigurationProperties(prefix = "student")
    public Student getStudent() {
        Student student = new Student();
        return student;
    }
}
```

(5) 编写测试类并运行,运行结果如图 2-5 所示。

在类中使用@Configuration 注解 Spring 的配置类时,使用@Configuration 注解的类相当于 XML 配置的< beans >元素,该类中的方法使用@Bean 注解注册组件,相当于 XML 配置中的< bean >元素。

@Bean 注解的方法不能是 private 或 final,注册 Bean 的 id 就是方法名,@Bean 注解虽然也可以用在@Component 注解的方法或其他的普通方法中,但是使用在@Configuration 注解类中是更为常见和推荐的用法。在非@Configuration 注解类的@Bean 注解方法中不能定义 Bean 间的依赖关系,如果定义在非@Configuration 注解类的依赖关系中,则有可能被当作一般的方法被调用,而不是用来作为 Bean 定义的方法。

2.4 多环境配置

在开发 Spring Boot 应用时,通常同一套程序会被应用和安装到几个不同的环境,如开发、测试、生产等。其中每个环境的数据库地址、服务器端口等配置都会不同,如果在为不同环境打包时都要频繁修改配置文件,那必将是个非常烦琐且容易发生错误的事。

2.4.1 使用 Profile 进行多环境配置

Spring Boot 为环境配置提供了非常好的支持,即多环境配置。可以将多种环境的参数一起配置到项目中,只需要执行不同的运行命名,就可以达到切换环境的目的。

接下来通过示例讲解多环境配置文件的用法。

(1) 新建一个名为 springboot0206 的项目,选择 Web 依赖。

(2) 编写多环境配置文件。

在 Spring Boot 中,多环境配置文件名必须满足 application-{profile}.properties 或 application-{profile}.yaml 的固定格式,其中{profile}对应环境标识。如图 2-7 是项目的 3 个配置文件,其中,application.yaml 是启动服务时自动加载的配置文件,application-dev.yaml 是开发环境的配置文件,application-test.yaml 是测试环境的配置文件。后两个文件在启动服务时,服务器不会自动加载。

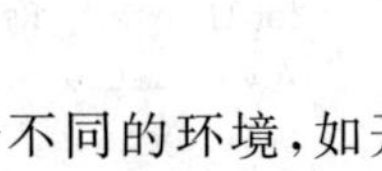

图 2-7 配置文件

(3) 添加环境参数。

针对各环境在不同的配置文件 application-dev.yaml 和 application-test.yaml 中设置不同的 server.port 属性,如 dev 环境设置为 1111、test 环境设置为 2222,代码如下。

① application-dev.yaml。

```yaml
server:
    port: 1111
  data:
    test:
      env-name: dev
      envconfig: 127.0.0.1:1111
```

② application-test.yaml。

```yaml
server:
    port: 2222
  data:
    test:
      env-name: test
      envconfig: 127.0.0.1:2222
```

③ 读取配置参数。

```java
@Component
@ConfigurationProperties(prefix = "data.test")
public class DataConfig {
  private String envName;
  private String envconfig;
  //省略 set 方法和 get 方法
}
```

④ 验证环境参数。

```java
@RestController
public class HomeController {
    @Autowired
    private DataConfig config;
    @RequestMapping("/env")
    public Object testEnv() {
        return config;
    }
}
```

⑤ 设置默认环境文件。

application.yaml 文件通过 spring.profiles.active 来具体激活一个或者多个配置文件。

```yaml
spring:
    profiles:
        active: test
```

此时 Spring Boot 就会加载 application-test.yaml 配置文件内容。

⑥ 运行程序，查看效果，如图 2-8 所示。

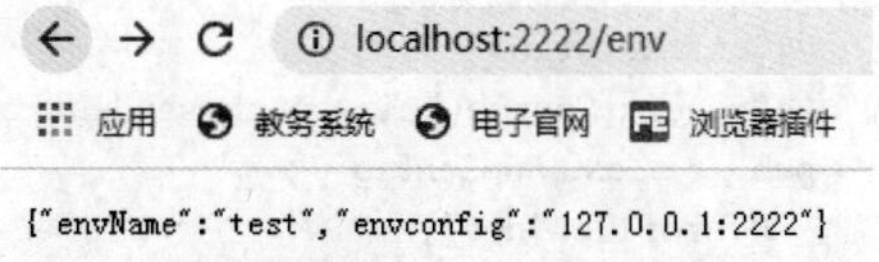

图 2-8 使用 Profile 文件多环境配置的运行结果

2.4.2 使用@Profile 进行多环境配置

除了使用 Profile 文件进行多环境配置外，还可以使用@Profile 注解进行多环境配置。在不同的场景下，@Profile 注解给出不同的类实例。例如，在生产环境中给出的 DataSource 实例与在测试环境中给出的 DataSource 实例是不同的。

@Profile 一般在@Configuration 下使用，标注在类或者方法上，标注时填入一个字符串（如"dev"），作为一个场景或区分。在 Spring 中，配置 Profile 为 dev 时，就可以获取该 Bean 了。

接下来通过示例讲解@Profile 多环境配置文件的用法。

（1）新建一个名为 springboot0207 的项目，选择 Web 依赖。

（2）在 it.com.boot.springboot0207.po 包下编写 Student 类和 Birthday 类。

（3）添加环境参数。

新建 student.properties 文件，针对不同的环境配置各自的学生信息，具体代码如下。

```
#开发环境
dev.stuName = 张三
dev.stuAge = 26
dev.stuSex = 男
dev.address = 天津,北京,上海
dev.score.math = 90
dev.score.english = 80
dev.score.computer = 95
dev.birthday.year = 1999
dev.birthday.month = 1
dev.birthday.day = 2
#测试环境
test.stuName = 李四
test.stuAge = 30
test.stuSex = 女
test.address = 重庆,浙江,江苏
test.score.math = 80
test.score.english = 90
test.score.computer = 70
test.birthday.year = 2000
test.birthday.month = 11
test.birthday.day = 12
```

（4）读取配置参数。

基于@Configuration 标注配置类，在配置里使用@Profile，分别表示哪个对象作用于哪个环境配置。

```
@Configuration
@PropertySource("classpath:student.properties")
public class AppConfig {
  @Profile("dev")
  @Bean
  @ConfigurationProperties(prefix = "dev")
  public Student getDevStudent() {
    Student student = new Student();
    return student;
  }
  @Profile("test")
  @Bean
  @ConfigurationProperties(prefix = "test")
  public Student getTestStudent() {
```

```
        Student student = new Student();
        return student;
    }
}
```

(5) 设置默认环境文件。

在全局配置文件 application.properties 中设置 spring.profiles.active=dev,激活使用 @Profile 注解构造的多环境配置。

(6) 运行程序,查看效果,如图 2-5 所示。

本章小结

本章主要讲解了 Spring Boot 的核心配置与注解,包括全局配置的使用、配置文件属性值的注入、Spring Boot 自定义配置和多环境配置等。希望通过本章的学习,大家能够掌握 Spring Boot 的核心配置,并灵活运用 Spring Boot 注解进行开发。

习题

在线测试

一、单选题

1. 以下关于 Spring Boot 配置文件属性注入的说法,正确的是(   )。
   A. 使用@ConfigurationProperties 注解可以批量注入各种类型属性值
   B. 使用@Value 注解需要逐个注入各种类型属性值
   C. 使用@ConfigurationProperties 和@Value 注解注入属性值需要设置属性 Set 方法
   D. 以上说法都错误

2. 以下关于 Spring Boot 多环境配置文件名格式的说法,正确的是(   )。
   A. application-dev.properties
   B. application.test.properties
   C. application.prod.yaml
   D. application_prod.yml

3. 以下关于 application.yaml 格式配置文件格式的说法,正确的是(   )。
   A. application.yaml 文件使用"key:value"格式配置属性,使用缩进控制层级关系
   B. 使用 YAML 文件的行内式写法定义数组时,包含属性值的中括号"[]"可以省略
   C. 使用 YAML 文件的行内式写法定义集合时,包含属性值的中括号"{}"可以省略
   D. 以上说法都错误

4. 使用命令行的方式激活 Spring Boot 多环境配置文件 application-dev.properties,以下指令正确的是(   )。
   A. java -jar xxx.jar -spring.profiles.active=dev
   B. java -jar xxx.jar --spring.profiles.active=dev
   C. java -jar xxx.jar --spring.profiles.active=application-dev.properties
   D. java -jar xxx.jar --spring.profiles.active=dev

二、多选题

1. 以下关于 Spring Boot 配置文件的说法,正确的是(　　)。
   A. Spring Boot 默认无法识别自定义 XML 配置文件
   B. Spring Boot 中通常使用@Configuration 注解配置类进行文件配置
   C. 定义@Configuration 配置类中 Bean 组件时必须使用 value 指定组件名称
   D. @Bean 属性通常放在配置类方法上

2. 以下关于 Spring Boot 配置文件属性注入的说法,正确的是(　　)。
   A. 注解@Value 注解是 Spring Boot 框架提供的
   B. 使用@ConfigurationProperties 注解进行配置文件属性值读取注入时,必须为每一个属性设置 setXX()方法
   C. 使用@ConfigurationProperties 注解进行配置文件属性值读取注入时,实体类必须为 Spring Bean 组件
   D. 以上说法都正确

3. 以下关于 application.yaml 格式配置文件属性配置的说法,错误的是(　　)。
   A. server.port:8081
   B. server:port:8081
   C. person:hobby:play read
   D. person:hobby:play,read,sleep

三、填空题

1. 使用@ConfigurationProperties 注解注入属性值,可以添加 Spring Boot 提供的配置处理器依赖_____。

2. 在 Spring Boot 配置类中可以在类中的方法上使用_____注解进行组件配置。

3. Spring Boot 提供了_____和_____两种主要方式进行多环境文件配置。

4. 依赖于 Spring Boot 中_____的思想,Spring Boot 已经对一些基本的配置进行了默认设置。

四、简答题

结合实际开发情况,简述@ConfigurationProperties 和@Value 两种注解的使用选择。

第3章 Spring Boot视图技术

本章学习目标
- 掌握 Thymeleaf 模板引擎。
- 掌握 Spring Boot 处理 JSON 数据。
- 掌握 Spring Boot 处理国际化问题。
- 掌握 Spring Boot 文件的上传和下载。
- 掌握 Spring Boot 的异常处理。

根据党的二十大报告，必须坚持"创新是第一动力"。创新是一个国家、一个民族发展进步的不竭动力，是推动人类社会进步的重要力量。

Web 应用开发是现代软件开发中的重要部分。Spring Boot 的 Web 开发内嵌了 Servlet 和服务器，并结合 Spring MVC 来完成开发。Web 开发是一种基于 B/S 架构的应用软件开发技术，分为前端和后端。前端的可视化及用户交互由浏览器实现，即以浏览器作为客户端，实现客户端与服务器远程的数据交互。

本章通过学习基于用户信息的 Web 层技术、集成 Spring MVC 技术和集成静态页面模板技术，学习创新的思维模式，并深刻体会这种时代宝贵的工匠精神。

3.1 创建静态 Web 页面

视频讲解

Spring Boot 项目在不使用任何模板引擎的前提下，想要直接访问 HTML 页面，是如何实现的呢？只需要把静态文件（HTML,JS,CSS）放在 resource 的 static 文件夹里面即可正常访问。下面通过示例讲解 Spring Boot 如何访问静态 Web 页面。

(1) 创建 Spring Boot Web 应用 springboot0301。

(2) 在项目的 src/main/resource/static 目录下创建 CSS 文件夹和 js 文件，将文件放入对应的文件夹下。

(3) 选中 static 目录并右击，选择新建 HTML 文件，命名为 index.html。HTML 文件代码如下：

```
<!DOCTYPE html>
<html lang = "zh_CN" xmlns:th = "http://www.thymeleaf.org">
<head>
    <meta charset = "UTF - 8">
```

```html
        <title>Title</title>
        <link href="css/style.css" rel="stylesheet" type="text/css">
        <script src="js/check.js"></script>
</head>
<body>
<div align="center"><br>
        <a class="title">用户登录</a><br>
</div>
<form name="form1" method="post">
        <table width="398" height="215" border="1" align="center" cellpadding="0" cellspacing="0">
                <tr>
                        <td width="394" height="213">
                                <table width="91%" height="80%" border="0" align="center" cellpadding="1" cellspacing="1">
                                        <tr>
                                                <td width="120" align="right">用户名:</td>
                                                <td width="208">
                                                        <input name="userid" type="text" id="userid" size="15" maxlength="20">
                                                </td>
                                        </tr>
                                        <tr>
                                                <td width="120" align="right">密码:</td>
                                                <td width="208">
                                                        <input name="password" type="password" id="password" size="15" maxlength="20">
                                                </td>
                                        </tr>
                                        <tr>
                                                <td width="120" height="23" align="right"> </td>
                                                <td width="208">
                                                        <div align="left">
                                                                <input type="submit" name="Submit" value="登录" onclick="javascript:return(checkform());">
                                                                <input type="reset" name="reset" value="重填">
                                                        </div>
                                                </td>
                                        </tr>
                                </table>
                                <div align="center"></div>
                                <div align="center"></div>
                                <div align="center"></div>
                                <div align="center"></div>
                        </td>
                </tr>
        </table>
</form>
</body>
</html>
```

（4）运行程序。

index.html 是默认的启动页面,在浏览器地址栏中输入 http://localhost:8080/后,结果如图 3-1 所示。

图 3-1　静态 Web 页面的运行效果图

3.2　Spring Boot 对 JSP 的支持

视频讲解

Spring Boot 默认不支持 JSP，因为 JSP 的性能相对较低。官方推荐使用 Thymeleaf，如果想在项目中使用 JSP，需要进行相关初始化工作。

下面通过实例演示 Spring Boot 如何访问 JSP 页面。

(1) 创建 Spring Boot Web 应用 springboot0302。由于 Spring Boot 使用 JSP 时需打包为 WAR 类型，所以在创建项目时修改打包方式，或者在后面的 pom.xml 文件中进行修改。

(2) 自动生成 ServletInitializer 类。

打开该类不难发现它继承了 SpringBootServletInitializer 这个父类，而 SpringBootServletInitializer 类是 Spring Boot 提供的 Web 程序初始化的入口，在使用外部容器(如使用外部 Tomcat)运行项目时会自动加载并且装配。

实现 SpringBootServletInitializer 的子类需要重写一个 configure 方法，方法内自动根据 Springboot0302Application.class 的类型创建一个 SpringApplicationBuilder，将其交付给 Spring Boot 框架来完成初始化运行配置。

(3) 整理脚本样式静态文件。

JS 脚本、CSS 样式、Image 等静态文件默认放置在项目的 src/main/resource/static 目录下。

(4) 配置 Spring Boot 支持 JSP。

打开 pom.xml 文件(Maven 配置文件)可以看到，之前构建项目时已经添加了 Web 模块。因为在 JSP 页面中使用 EL 和 JSTL 标签显示数据，所以除了添加 Servlet 和 Tomcat 依赖外，还需要添加 JSTL 依赖，具体代码如下。

```
<!-- WAR 包部署到外部的 Tomcat 中已经包含了以下依赖，在此需要对其进行添加否则会与内嵌
Tomcat 容器发生冲突 -->
<dependency>
    <groupId>org.springframework.boot</groupId>
```

```xml
    <artifactId>spring-boot-starter-tomcat</artifactId>
    <scope>provided</scope>
</dependency>
<!-- Servlet 依赖 -->
<dependency>
    <groupId>javax.servlet</groupId>
    <artifactId>javax.servlet-api</artifactId>
</dependency>
<!-- 配置 JSP JSTL 的支持 -->
<dependency>
    <groupId>javax.servlet</groupId>
    <artifactId>jstl</artifactId>
</dependency>
<!-- 引入 Spring Boot 内嵌的 Tomcat 对 JSP 的解析包,否则无法解析 JSP 页面 -->
<dependency>
    <groupId>org.apache.tomcat.embed</groupId>
    <artifactId>tomcat-embed-jasper</artifactId>
</dependency>
```

(5) 配置视图。

必须按照 JSP 文件之前的文件目录风格配置视图,因此需要在 src/main 目录下创建 webapp/WEB-INF/views 文件夹。修改 application.properties 文件,使得 Spring MVC 支持视图的跳转目录指向为 /main/webapp/views,代码如下。

```
spring.mvc.view.prefix=/WEB-INF/views/
spring.mvc.view.suffix=.jsp
```

(6) 创建 Book 实体类、BookController 控制器及对应的 BookDAO 数据层等,具体代码见源文档。

(7) 在 WEB-INF/views 目录下新建 booklist.jsp 文件,主要展示所有图书的信息,具体代码如下。

```jsp
<%@ taglib prefix="c" uri="http://java.sun.com/jsp/jstl/core" %>
<%@ page language="java" contentType="text/html; charset=UTF-8" pageEncoding="UTF-8" %>
<!DOCTYPE html PUBLIC "-//W3C//DTD HTML 4.01 Transitional//EN" "http://www.w3.org/TR/html4/loose.dtd">
<html>
<head>
<title>网上书店</title>
<link href="css/style.css" rel="stylesheet" type="text/css">
</head>
<body>
<%@ include file="header.jsp" %>
<br><a class="title">在线购书</a><br>
<table width="700" border="1" cellpadding="0" cellspacing="0">
    <tr>
        <td width="43%" height="26"><div align="center">书名</div></td>
        <td width="17%"><div align="center">作者</div></td>
        <td width="17%"><div align="center">出版社</div></td>
        <td width="5%"><div align="center">价格</div></td>
        <td width="9%"><div align="center">操作</div></td>
    </tr>
```

```
   <c:forEach items = "${bookList}" var = "book">
     <tr>
       <td height = "32"><div align = "center">${book.getTitle()}</div></td>
       <td><div align = "center">${book.getAuthor()}</div></td>
       <td><div align = "center">${book.getPublisher()}</div></td>
       <td><div align = "center">${book.getPrice()}</div></td>
       <td><div align = "center"><a href = "bookdetail.jsp?isbn = ${book.get综合业务数字网()}">
<img src = "images/buy.gif" width = "45" height = "16" border = "0"></a></div></td>
     </tr>
   </c:forEach>
</table>
<p> </p>
<%@ include file = "footer.jsp" %>
</body>
</html>
```

(8) 启动测试。

在浏览器地址栏中输入 http://localhost:8080/ 后, 在页面中输入任意用户名和密码, 单击登录按钮, 就可以进入网上书店的首页, 如图 3-2 所示。

图 3-2　JSP 页面的运行效果图

3.3　Thymeleaf 的基本语法

Thymeleaf 是一款用于渲染 XML/XHTML/HTML5 内容的模板引擎。与 JSP、Velocity、FreeMaker 等类似, 它也可以轻易地与 Spring MVC 等 Web 框架进行集成, 以作为 Web 应用的模板引擎。与其他模板引擎相比, Thymeleaf 最大的特点是能够直接在浏览器中打开并正确显示模板页面, 而不需要启动整个 Web 应用。

Thymeleaf 支持 Spring Expression Language 语言作为方言, 也就是 SpEL, SpEL 是可

以用于 Spring 中的一种 EL 表达式。

简而言之，与以往使用过的 JSP 不同，Thymeleaf 使用 HTML 的标签来完成逻辑和数据的传入及渲染，而不用像 JSP 一样作为一个 Servlet 被编译再生成。即便是单独用 Thymeleaf 开发的 HTML 文件，依旧可以正确打开并有少量（相对）有价值的信息，并且是可以被浏览器直接打开的。用 Thymeleaf 完全替代 JSP 是可行的。

Thymeleaf 的主要作用是把 model 中的数据渲染到 HTML 中，因此其语法主要是如何解析 model 中的数据。在 HTML 页面中引入 Thymeleaf 命名空间，即在 HTML 模板文件中对动态的属性使用 th: 命名空间修饰。

```
< html lang = "en" xmlns:th = "http://www.thymeleaf.org">
```

这样才可以在其他标签里面使用 th:* 这样的语法，这是下面语法的前提。

3.3.1 变量表达式

通过项目 springboot0303 来学习 Thymeleaf 的语法。

（1）新建一个实体类：User

```
public class User {
    private String name;
    private Integer age;
    private String sex;
}
```

（2）在模型中添加数据：

```
@GetMapping("show1")
public String show2(Model model){
    User user = new User();
    user.setAge(21);
    user.setName("张三");
    user.setSex("男");
    model.addAttribute("user", user);
    return "index";
}
```

语法说明：

Thymeleaf 通过 ${} 来获取 model 中的变量，注意这不是 EL 表达式，而是 OGNL 表达式。

示例：

在页面获取 user 数据，其效果如图 3-3 所示。

```
< h1 >
    欢迎您:< span th:text = " ${user.name}">请登录</ span>
</ h1 >
```

OGNL 表达式的语法与 EL 表达式几乎是一样的。两者的区别在于，OGNL 表达式写在一个名为 th:text 的标签属性中，该标签即为指令。

图 3-3 利用 OGNL 表达式获取数据

Thymeleaf 崇尚自然模板,即模板是纯正的 HTML 代码,脱离模板引擎,在纯静态环境也可以直接运行。如果直接在 HTML 中编写 ${}这样的表达式,显然在静态环境下就会出错,这不符合 Thymeleaf 的理念。

Thymeleaf 中所有的表达式都需要写在指令中,指令是 HTML5 中的自定义属性,在 Thymeleaf 中的所有指令都是以 th:开头的。由于表达式 ${user.name}是写在自定义属性中的,因此在静态环境下,表达式的内容会被当作是普通字符串,浏览器会自动忽略这些指令,从而不会报错。

现在不经过 Spring MVC,而是直接用浏览器打开页面,如图 3-4 所示。

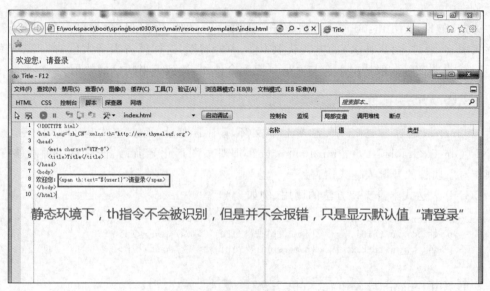

图 3-4　用浏览器打开页面

在静态环境下,th 指令不会被识别,但是浏览器也不会报错,而是把它当作一个普通属性处理。这样,span 的默认值"请登录"就会显示在页面上。如果是在 Thymeleaf 环境下,th 指令就会被识别和解析,而 th:text 的含义就是替换所在标签中的文本内容,于是 user.name 的值就替代了 span 中默认的"请登录"。

指令的设计,正是 Thymeleaf 的高明之处,也是它优于其他模板引擎的原因。动静结合的设计,使得无论是前端开发还是后端开发都可以完美契合。

需要注意的是,如果不支持这种 th:的命名空间写法,那么可以把 th:text 换成 data-th-text,Thymeleaf 同样可以兼容。

另外,出于安全考虑,th:text 指令会把表达式读取到的值进行处理,防止 HTML 的注入。例如,将字符串"你好"存入到 Model 中,在页面中可以使用如下两种方式获得消息文本:

```
< span th:text = " ${msg}"></span> <!-- 不对 HTML 标签解析,输出原始内容 -->
< span th:utext = " ${msg}"></span> <!-- 对 HTML 标签解析,输出加粗的"你好" -->
```

3.3.2　自定义变量

下面这个例子是分别展示获取到的用户的所有信息。

```
< h2 >
    < p >姓名：< span th:text = " ${user.name}">Jack </ span ></ p >
    < p >年龄：< span th:text = " ${user.age}"> 21 </ span ></ p >
</ h2 >
```

当数据量比较多时，频繁地写 user. 就会非常麻烦。因此，Thymeleaf 提供了自定义变量来解决该问题。

```
< h2 th:object = " ${user}">
    < p >姓名：< span th:text = " *{name}">Jack </ span ></ p >
    < p >年龄：< span th:text = " *{age}"> 21 </ span ></ p >
</ h2 >
```

首先，在 h2 上用 th:object = " ${user}"获取 user 的值，并且保存。然后，在 h2 内部的任意元素上，通过 *{属性名}的方式获取 user 中的属性，这样就省去了大量的 user. 前缀。th:object 声明变量一般情况下会与 *{}一起配合使用，达到事半功倍的效果。

3.3.3 方法

Thymeleaf 通过 ${}来获取容器上下文环境中的变量，而表达式是 OGNL 表达式。OGNL(Object-Graph Navigation Language)，即对象图形化导航语言，它是一种能够方便地操作对象属性的开源表达式语言。

OGNL 表达式本身支持方法的调用，例如：

```
< h2 th:object = " ${user}">
    < p >姓：< span th:text = " *{name.split(' ')[0]}">张</ span ></ p >
    < p >名：< span th:text = " *{name.split(' ')[1]}">三</ span ></ p >
</ h2 >
```

这里调用了 name(字符串)的 split 方法。

Thymeleaf 提供了一些内置对象，并且在这些对象中提供了一些方法，以方便调用。在获取这些对象后，需要使用"#对象名"来引用。

(1) 一些环境相关对象。

#ctx：获取 Thymeleaf 自己的 Context 对象。

#request：如果是 Web 程序，可以获取 HttpServletRequest 对象。

#response：如果是 Web 程序，可以获取 HttpServletResponse 对象。

#session：如果是 Web 程序，可以获取 HttpSession 对象。

#servletContext：如果是 Web 程序，可以获取 HttpServletContext 对象。

(2) Thymeleaf 提供的全局对象。

#dates：处理 java.util.date 的工具对象。

#calendars：处理 java.util.calendar 的工具对象。

#numbers：用来对数字格式化的方法。

#strings：用来处理字符串的方法。

#bools：用来判断布尔值的方法。

#arrays：用来处理数组的方法。

#lists：用来处理 List 集合的方法。

#sets：用来处理 Set 集合的方法。

#maps：用来处理 Map 集合的方法。

例如，在环境变量中添加日期类型对象：

```
@GetMapping("show2")
public String show2(Model model){
    model.addAttribute("today", new Date());
    return "index";
}
```

在页面中处理如下：

```
<p>
    今天是：<span th:text = "${#dates.format(today,'yyyy-MM-dd')}">2022-01-14</span>
</p>
```

添加日期对象的运行效果如图 3-5 所示。

图 3-5　#dates 全局对象

3.3.4　字面值

有时需要在指令中填写基本类型，如字符串、数值、布尔等，但并不希望被 Thymeleaf 解析为变量，这时将其称为字面值。

（1）字符串字面值。

使用一对单引号引用的内容就是字符串字面值。

`<p>你正在观看<span th:text = "'Thymeleaf'">template</span>的字符串常量值.</p>`

th:text 指令中的 Thymeleaf 并不会被认为是变量，而会被认为是一个字符串。字符串字面值的程序运行效果如图 3-6 所示。

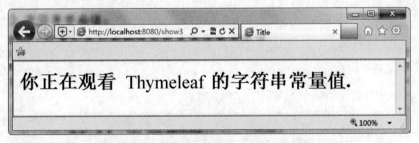

图 3-6　字符串字面值的程序运行效果图

（2）数字字面值。

数字字面值不需要任何特殊语法，直接写入内容即可，而且可以进行算术运算。

`<p>今年是<span th:text = "2022"></span>.</p>`
`<p>两年后将会是<span th:text = "2022 + 2"></span>.</p>`

数字字面值的程序运行效果如图 3-7 所示。

（3）布尔字面值。

布尔类型的字面值是 true 或 false。

```
<div th:if = "true">
    你填的是 true
</div>
```

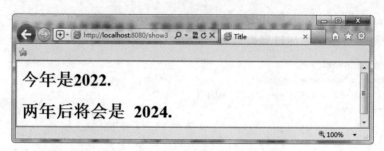

图 3-7　数字字面值的程序运行效果图

这里引用了一个 th:if 指令。

3.3.5　拼接

在指令中经常会遇到普通字符串与表达式拼接的情况：

`<span th:text="'欢迎您:' + ${user.name}"></span>`

字符串字面值需要用单引号，拼接起来非常麻烦，Thymeleaf 对此进行了简化，使用一对 | 即可。

`<span th:text="|欢迎您:${user.name}|"></span>`

这与上面的语法是完全等效的，同时省去了字符串字面值的书写。

3.3.6　运算

需要注意的是，${} 内部通过 OGNL 表达式引擎进行解析，而外部通过 Thymeleaf 的引擎进行解析，因此运算符尽量放在 ${} 外。

（1）算术运算。

支持的算术运算符：＋、－、＊、/、％。

`<span th:text="${user.age}"></span><span th:text="${user.age} % 2 == 0"></span>`

算术运算符的程序运行效果如图 3-8 所示。

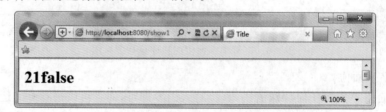

图 3-8　算术运行符的程序运行效果图

（2）比较运算。

支持的比较运算：＞、＜、＞＝、＜＝、＝＝、！＝。需要注意的是，＝＝和！＝不仅可以比较数值，而且可以比较对象，与 equals 的功能类似。

除了使用常规的比较运算符外，还可以使用别名：gt(＞)、lt(＜)、ge(＞＝)、le(＜＝)、not(！)、eq(＝＝)、ne(！＝)。

`<span th:text="'你的年龄是否小于30:' + (${user.age} < 30)"></span>`
`<span th:text="'你的年龄是否超过30:' + (${user.age} gt 30)"></span>`

比较运算符的程序运行效果如图 3-9 所示。

图 3-9　比较运算符的程序运行效果图

（3）三元运算。

三元运算符的表达式为 condition ? then :else。

condition：条件。

then：条件成立的结果。

else：不成立的结果。

其中的每个部分都可以是 Thymeleaf 中的任意表达式。

< span th:text = "'你的性别：' + (${user.sex} ? '男':'女')"></span>

三元运算符的程序运行效果如图 3-10 所示。

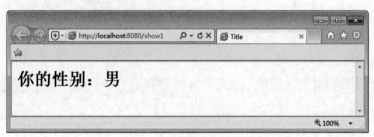

图 3-10　三元运算符的程序运行效果图

（4）默认值。

指令中的取值有时可能为空，这时需要进行非空判断，可以使用表达式"?:默认值"简写。

< span th:text = "'你的年龄是' + (${user.age} ?: 20)"></span>

当前面的表达式值为 null 时，就会使用后面的默认值。

注意："?:"之间没有空格。

默认值的程序运行效果如图 3-11 所示。

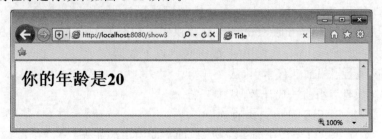

图 3-11　默认值的程序运行效果图

3.3.7 循环

当信息页面的数据格式相同时，页面通常对它们进行循环迭代。JSTL 有一个 <c:foreach>，同理 Thymeleaf 也有一个 th:each，其作用都是一样的，用于遍历数组、List、Set、Map 等数据。

假如有用户的 java.util.List 集合的 users 在 Context 中。

```
<tr th:each="user : ${users}">
    <td th:text="${user.name}"></td>
    <td th:text="${user.age}"></td>
    <td th:text="${user.sex}"></td>
</tr>
```

java.util.List 类型不是可以在 Thymeleaf 中使用迭代的唯一值类型，下面这些类型的对象都可以通过 th:each 进行迭代。

（1）任何实现 java.util.Iterable 接口的对象，其值将被迭代器返回，而不需要在内存中缓存所有值。

（2）任何实现 java.util.Enumeration 接口的对象。

（3）任何实现 java.util.Map 接口的对象，在迭代映射时，iter 变量将是 java.util.Map.Entry 类。

（4）任何数组。

（5）任何将被视为包含对象本身的单值列表。

循环语句的程序运行效果如图 3-12 所示。

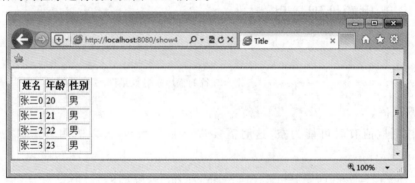

图 3-12　循环语句的程序运行效果图

在迭代过程中，经常会使用到它的一些迭代状态，如当前迭代的索引、迭代变量中的元素的总数、当前迭代的是奇数还是偶数、当前是否为第一个元素、当前是否为最后一个元素等。在 JSP 的 JSTL 中，<c:foreach items="${list}" var="li" varStatus="status">包含 status 属性。在使用 th:each 时，Thymeleaf 提供了一种用于跟踪迭代状态的机制：状态变量。状态变量在每个 th:each 属性中定义，并包含以下数据。

（1）index 属性：当前迭代索引，从 0 开始。

（2）count 属性：当前迭代计数，从 1 开始。

（3）size 属性：迭代变量中元素的总量。

（4）current 属性：每次迭代的 iter 变量，即当前遍历的元素。

(5) even/odd 布尔属性：当前迭代是偶数还是奇数。
(6) first 布尔属性：当前迭代是否为第一个迭代。
(7) last 布尔属性：当前迭代是否为最后一个迭代。

```
< tr th:each = "user ,loopStatus: ${users}" th:class = " ${loopStatus.odd}?odd" >
    < td th:text = " ${loopStatus.count}"></td>
    < td th:text = " ${user.name}"></td>
    < td th:text = " ${user.age}"></td>
     < td th:text = " ${user.sex}"></td>
</tr>
```

状态变量的程序运行结果如图 3-13 所示。

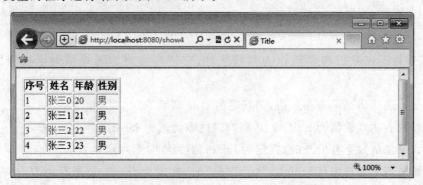

图 3-13　状态变量的程序运行结果图

迭代状态变量（本例中的 loopStatus）在 th:each 属性中，通过在变量 user 后直接写其名称来定义，用逗号分隔。与 iter 变量一样，状态变量的作用范围也是 th:each 属性的标签定义的代码片段中。

如果没有显式地设置状态变量，则 Thymeleaf 将始终创建一个默认的迭代变量，该状态迭代变量名称为：迭代变量＋"Stat"。

```
< tr th:each = "user : ${users}" th:class = " ${userStat.odd}?odd" >
    < td th:text = " ${userStat.count}"></td>
    < td th:text = " ${user.name}"></td>
    < td th:text = " ${user.age}"></td>
    < td th:text = " ${user.sex}"></td>
</tr>
```

3.3.8　逻辑判断

很多时候只有在满足某个条件时，才将一个模板片段显示在结果中，否则不进行显示。例如，只有当用户年龄小于 18 岁时，才显示为未成年人，否则不显示。th:if 属性用于满足逻辑判断的需求。

```
< tr th:each = "user ,loopStatus: ${users}" th:class = " ${loopStatus.odd}?odd" >
    < td th:text = " ${loopStatus.count}"></td>
    < td th:text = " ${user.name}"></td>
    < td th:text = " ${user.age}"></td>
    < td th:text = " ${user.sex}"></td>
    < td >< span th:if = " ${user.age}< 18">未成年</span></td>
</tr>
```

逻辑判断的程序运行结果如图 3-14 所示。

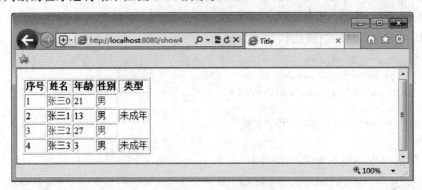

图 3-14　逻辑判断的程序运行效果图

th:if 属性不仅只以布尔值作为判断条件，它还将按照如下规则判定指定的表达式结果。

（1）如果表达式结果为布尔值，则判定为 true 或者 false。

（2）如果表达式的值为 null，th:if 将判定此表达式为 false。

（3）如果值是数字且为 0 时，判定为 false；不为零时则判定为 true。

（4）如果值是字符串且为"false""off""no"时，判定为 false；否则判定为 true。当字符串为空时，也判定为 true。

（5）如果值不是布尔值、数字、字符或字符串的其他对象，只要不为 null，则判定为 true。

th:unless 是 th:if 的反向属性，它们判断的规则一致，只是 if 当结果为 true 时进行显示，unless 当结果为 false 进行显示。

3.3.9　分支控制 switch

th:switch/th:case 与 Java 中的 switch 语句等效，有条件地显示匹配的内容。只要其中一个 th:case 的值为 true，则同一个 switch 语句中的其他 th:case 属性都将被视为 false。当有多个 case 的值为 true 时，则只取第一个。

switch 语句的 default 选项指定为 th:case ="*"，即当没有 case 的值为 true 时，将显示 default 的内容，如果有多个 default，则只取第一个。

```
< tr th:each = "user ,loopStatus: ${users}" th:class = " ${loopStatus.odd}?odd" >
    < td th:text = " ${loopStatus.count}"></td>
    < td th:text = " ${user.name}"></td>
    < td th:text = " ${user.age}"></td>
    < td th:text = " ${user.sex}"></td>
    < td th:switch = " ${user.age}< 18">
        < span th:case = "true">未成年</span>
        < span th:case = " * ">成年人</span>
    </td>
</tr>
```

分支语句程序的运行效果如图 3-15 所示。

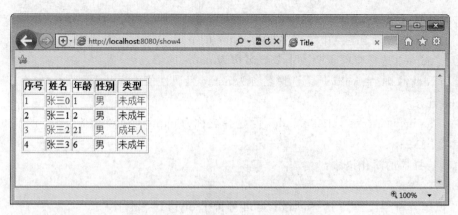

图 3-15 分支语句程序的运行效果图

3.3.10 Thymeleaf 模板片段

系统中的很多页面都有公共内容,如菜单、页脚等,这些公共内容可以提取并放在一个称为"模板片段"的公共页面中,其他页面可以引用该公共页面。

模板片段可以是 HTML 标签,也可以使用 th:fragment 属性定义片段。

首先定义一个页脚片段 footer.html,代码如下所示。

```
< span th:fragment = "frag1"> frag1 </span>
< span th:fragment = "frag2"> frag2 </span>
< div id = "footer1"> footer1 </div>
< div >
    < div id = "footer2"> footer2 </div>
</div>
< div >
    < span class = "content"> footer3 </span>
    < span class = "content"> footer4 </span>
</div>
< div th:fragment = "welcome(userName)">
    < span th:text = "|hello,| + ${userName}"></span>
</div>
```

接下来就可以在 Web 页面中使用 th:insert、th:replace、th:include 属性来包含页脚片段,这 3 个属性的区别如下。

- th:insert:在当前标签里面插入模板中的标签。
- th:replace:替换当前标签为模板中的标签。
- th:include:在标签里面插入模板的标签内容。

这里以使用 th:insert 属性插入片段为例,介绍其语法规则,其他两个属性类似。

(1) th:insert="~{模板名称}"。

插入模板的整个内容。

(2) th:insert="~{模板名称::选择器}"。

插入模板的指定内容,选择器可以对应 th:fragment 定义的名称,也可以用类似 JQuery 选择器的语法选择部分片段。

片段选择器语法：

① /name：选择子节点中节点名称为 name 的节点。

② //name：选择全部子节点中节点名称为 name 的节点。

③ name[@attr='value']：选择名称为 name 且属性值为 value 的节点，如有多个属性则用 and 连接。

④ //name[@attr='value'][index]：选择名称为 name 且属性值为 value 的节点，并指定节点索引。

片段选择器的简化语法：

① 可以省略@符号。

② 使用#符号代替 id 选择，如 div#id 等价于 div[id='id']。

③ 使用.符号代替 class 选择，如 div.class 等价于 div[class='class']。

④ 使用%代替片段引用，如片段节点使用了 th:ref 或 th:fragment，则可使用 div%ref 来选取节点。

(3) th:insert="~{::选择器}"。

不指定模板名称，则选择器作用于当前页面。

(4) th:insert="~{this::选择器}"。

与"~{::选择器}"类似，不同之处是在本页面找不到片段时，会到模板引擎的 process 方法处理的模板中寻找片段。

(5) 模板片段也支持传入变量，其引用语法为~{footer.html::名称(参数)}。

Thymeleaf 引用模板片段的代码如下所示。

```
<h4>th:insert引用片段</h4>
引用指定模板的整个内容
<div th:insert = "~{footer.html}"></div>
引用指定模板的片段
<div th:insert = "~{footer.html::frag1}"></div>
引用本页面的片段
<div th:insert = "~{::frag3}"></div>
<div th:insert = "~{this::frag3}"></div>
<div th:fragment = "frag3"> frag3 </div>


<h4>th:replace、th:include 与 th:insert 的区别</h4>
<div th:replace = "~{footer.html::frag1}"></div>
<div th:include = "~{footer.html::frag1}"></div>


<h4>片段选择器的部分用法</h4>
<div th:insert = "~{footer.html::div[@id = 'footer1']}"></div>
<div th:insert = "~{footer.html:://div#footer2}"></div>
<!-- <div th:insert = "~{footer.html::#footer2}"></div> -->
<div th:insert = "~{footer.html::span[class = 'content']}"></div>
<div th:insert = "~{footer.html:://span[class = 'content'][0]}"></div>
<div th:insert = "~{footer.html:://span.content}"></div>
<!-- <div th:insert = "~{footer.html::.content}"></div> -->
<!-- <div th:insert = "~{footer.html::.content[0]}"></div> -->
<div th:insert = "~{footer.html::span%frag1}"></div>
<!-- <div th:insert = "~{footer.html::%frag1}"></div> -->
<h4>含有变量的片段引用</h4>
```

```
<div th:insert = "~{footer.html::welcome('小明')}"></div>
```

IDEA 运行后,查看网页源代码,代码如下:

```
<h4> th:insert 引用片段</h4>
引用指定模板的整个内容
<div>
    <span> frag1 </span>
    <span> frag2 </span>
    <div id = "footer1"> footer1 </div>
    <div>
        <div id = "footer2"> footer2 </div>
    </div>
    <div>
        <span class = "content"> footer3 </span>
        <span class = "content"> footer4 </span>
    </div>
    <div>
        <span> hello,null </span>
    </div>
</div>
引用指定模板的片段
<div><span> frag1 </span></div>
引用本页面的片段
<div><div> frag3 </div></div>
<div><div> frag3 </div></div>
<div> frag3 </div>

<h4> th:replace、th:include 与 th:insert 的区别</h4>
<span> frag1 </span>
<div> frag1 </div>

<h4>片段选择器的部分用法</h4>
<div><div id = "footer1"> footer1 </div></div>
<div><div id = "footer2"> footer2 </div></div>
<div><span class = "content"> footer3 </span><span class = "content"> footer4 </span></div>
<div><span class = "content"> footer3 </span></div>
<div><span class = "content"> footer3 </span><span class = "content"> footer4 </span></div>
<div><span> frag1 </span></div>

<h4>含有变量的片段引用</h4>
<div>
    <div>
    <span> hello,小明</span>
    </div>
</div>
```

3.4 实现基于 Thymeleaf 的 Web 应用

视频讲解

3.3 节已经对 Thymeleaf 的基本语法及使用进行了介绍,接下来通过一个项目重点讲解 Spring Boot 与 Thymeleaf 模板引擎的整合使用。

(1) 创建 Spring Boot Web 应用 springboot0304。

（2）选择相应依赖。

使用 Spring Initializr 方式创建项目，并在 Dependencies 依赖选择中选择 Web 模块下的 Web 场景依赖和 Template Engines 模块下的 Thymeleaf 场景依赖，然后根据提示完成项目的创建。引入的场景依赖效果如图 3-16 所示。

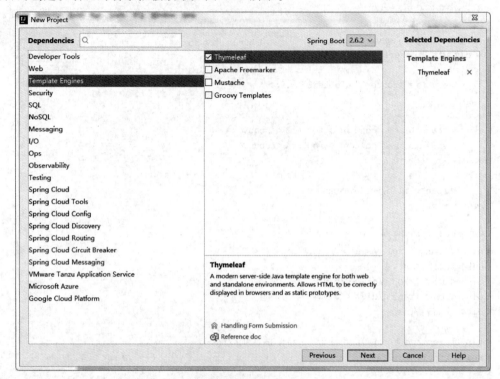

图 3-16　引入的场景依赖效果图

项目创建成功后，在 pom.xml 文件中就添加了 Thymeleaf 依赖，代码如下。

```xml
<dependency>
    <groupId>org.springframework.boot</groupId>
    <artifactId>spring-boot-starter-thymeleaf</artifactId>
</dependency>
```

（3）添加 Thymeleaf 配置。

在 application.properties 文件中，添加相关的配置项，代码如下。

```
#开发配置为false,避免修改模板后重启服务器
spring.thymeleaf.cache=false
#模板的模式,支持 HTML、XML、TEXT、JAVASCRIPT
spring.thymeleaf.mode=HTML5
#编码,可不用配置
spring.thymeleaf.encoding=utf-8
#内容类别,可不用配置
spring.thymeleaf.content-type=text/html
```

（4）整理脚本样式静态文件。

为了保证页面美观效果，需要在项目的静态资源文件夹 static 下创建相应的目录和文件，具体代码见源文档。

(5) 创建 Book 实体类、BookController 控制器及对应的 BookDAO 数据层等,具体代码见源文档。

(6) 在 resources/templates 目录下创建模板片段 header.html 和 footer.html,在此展示 header.html 代码。footer.html 代码比较简单,具体代码见源文档。

```html
    <div align="center" th:fragment="head">
    <table width="700" border="1" cellpadding="0" cellspacing="0" bordercolor="#CCCCCC">
        <tr>
            <td width="111" height="87" align="center"><img src="images/logo.gif" width="91" height="72"><br></td>
            <td width="509" align="center"><img src="images/banner.bmp" width="510" height="80"></td>
            <td align="center" width="70"><a href="">网站管理</a></td>
        </tr>
    </table>
    <table width="700" border="1" cellpadding="0" cellspacing="0" bordercolor="#CCCCCC">
        <tr>
            <td width="164"><div align="center"><a href="">首页</a></div></td>
            <td width="224"><div align="center"><a href="">在线购书</a></div></td>
            <td width="304" align="right"><div align="center"><a href="">我的购物车</a></div><div align="center"></div></td>
            <td width="304" align="right"><div align="center"><a href="">我的订单</a></div></td>
            <td width="304" align="right"><div align="center"><span th:text="'|当前日期:${dateString}|'"/></div></td>
        </tr>
    </table>
    </div>
```

(7) 创建模板页面 booklist.html 文件,并引入 head 和 foot 代码片段,主要展示所有图书的信息,具体代码如下。

```html
    <div th:insert="~{header.html::head}"></div>
    <div align="center"><a class="title">在线购书</a><br>
    <table width="700" border="1" cellpadding="0" cellspacing="0">
      <tr>
        <td width="43%" height="26"><div align="center">书名</div></td>
        <td width="17%"><div align="center">作者</div></td>
        <td width="17%"><div align="center">出版社</div></td>
        <td width="5%"><div align="center">价格</div></td>
        <td width="9%"><div align="center">操作</div></td>
    </tr>
    <div th:each="book:${bookList}">
        <tr>
            <td height="32"><div align="center"><span th:text="${book.getTitle()}"></span></div></td>
            <td><div align="center"><span th:text="${book.getAuthor()}"></span></div></td>
            <td><div align="center"><span th:text="${book.getPublisher()}"></span></div></td>
            <td><div align="center"><span th:text="${book.getPrice()}"></span></div></td>
            <td><div align="center"><a href=""><img src="images/buy.gif" width="45" height="16" border="0"></a></div></td>
        </tr>
    </div>
```

```
        </table>
    </div>
    <div th:replace = "~{footer.html::foot}"></div>
```

(8) 启动测试。

在浏览器地址栏中输入 http://localhost:8080/login 后,就可以进入网上书店的首页,如图 3-17 所示。

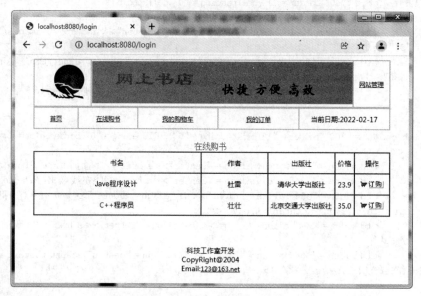

图 3-17 运行代码片段效果图

3.5 Spring Boot 中的页面国际化实现

什么是国际化?例如,dubbo.apache.org 是一个默认英文的网站,单击右上角的中文就会切换成中文网站,这就是国际化。

在 Spring Boot 的 Web 应用中实现页面信息国际化非常简单,下面通过示例讲解国际化的实现过程。

(1) 创建 Spring Boot Web 应用 springboot0305。

(2) 选择 Web 场景依赖和 Thymeleaf 场景依赖。

(3) 编写国际化配置文件。

① 在 resources 资源文件下新建一个 i18n 目录,用于存放国际化配置文件。选中此文件夹并右击,在弹出的命令选择器中选择 New→Resource Bundle 命令,如图 3-18 所示。

Resource Bundle 是一堆前缀名称相同但后缀名称不同的属性文件的集合,且至少包含两个有着相似前缀名称的属性文件,如 book_en_US.properties 和 book_zh_CN.properties。Resource Bundle 从字面上理解其实就是资源包,为了方便统一管理繁多的国际化文件,只是在 IntelliJ IDEA 内显示上多了一层名为 Resources 的 Resource Bundle 目录,但在实际物理目录下的 book_*.properties 等文件仍在 i18n 目录下。

第3章 Spring Boot视图技术

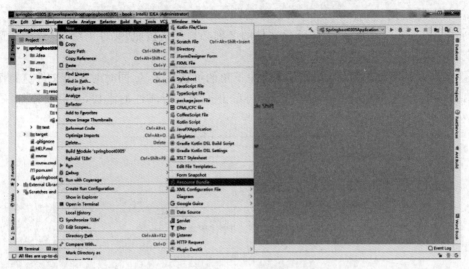

图 3-18　选择 Resource Bundle 命令

② 选择 Resource Bundle 后会弹出如图 3-19 所示的窗口,在窗口中填写 Resource Bundle 的基础名称 book,勾选 User XML-based properties files 则会创建 XML 格式的属性文件。Project locale 表示项目里已经存在的区域,Locales to add 表示添加相应的区域,单击右边的＋号即可添加,多个区域用英文的逗号隔开,如图 3-20 所示。

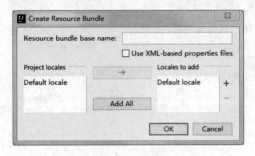

图 3-19　新建 Resource Bundle

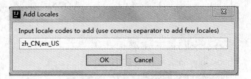

图 3-20　添加 Locales

③ 区域添加完成后,可以在 Locales to add 看到已经添加的区域,如图 3-21 所示。单击 OK 按钮,生成 Resource Bundle,如图 3-22 所示。

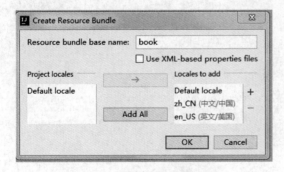

图 3-21　添加完成区域效果

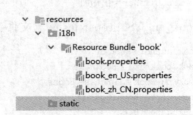

图 3-22　生成 Resource Bundle

book.properties 为自定义默认语言配置文件,book_zh_CN.properties 为自定义中文国际化文件,book_en_US.properties 为自定义英文国际化文件。

需要说明的是,Spring Boot 默认识别的语言配置文件为类路径 resources 下的 messages.properties,其他语言国际化文件的名称必须严格按照"文件前缀名_语言代码_国家代码.properties"的形式命名。

④ 点进3个文件中的任意一个,单击左下角的可视化工具,同时对3个文件进行编写,如图3-23所示。先单击 Resource Bundle 再单击＋号进行属性的添加,只需要添加一个即可,另外两个会自动加上。

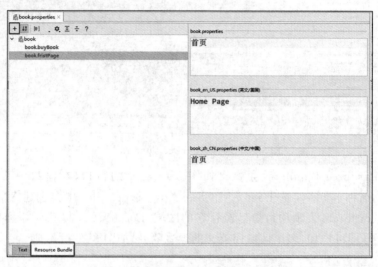

图 3-23　利用可视化工具添加属性

⑤ 查看配置文件。

book.properties 和 book_zh_CN.properties 配置文件:

book.author = 作者
book.bookname = 书名
book.buyBook = 在线购书
book.cart = 购物车
book.date = 当前日期
book.development = 科技工作室开发
book.fristPage = 首页
book.management = 网站管理
book.operation = 操作
book.order = 订单
book.price = 价格
book.publisher = 出版社

book_en_US.properties 配置文件:

book.author = Author
book.bookname = BookName
book.buyBook = Purchase Books
book.cart = hopping Cart
book.date = Date
book.development = Technology Studio Development
book.fristPage = Home Page
book.management = Website Management
book.operation = Operation

```
book.order = Order
book.price = Price
book.publisher = Publisher
```

（4）启用国际化配置。

Spring Boot 对国际化的自动配置涉及 MessageSourceAutoConfiguration 类的方法，这里 Spring Boot 已经自动配置好了管理国际化资源文件的组件 ResourceBundleMessageSource。资源文件 messages 存放在 i18n 目录下，所以要配置 messages 的路径。在 application.properties 文件里配置国际化信息，代码如下。

```
# 关闭 Thymeleaf 模板缓存
spring.thymeleaf.cache = false
# 绑定国际化配置文件
spring.messages.basename = i18n.book
```

注意：如果配置文件可以直接放在类路径下的 messages.properties 中，则不用将配置文件再在 application.properties 里配置。

（5）以 springboot0304 为基础，创建相应的 JS、CSS、图片、对应的类和页面。修改 head.html 页面，并获得国际化信息，代码如下。

```
<table width = "700" border = "1" cellpadding = "0" cellspacing = "0" bordercolor = "#CCCCCC">
    <tr align = "center">
        <td width = "164"><a href = "">[[#{book.firstPage}]]</a></td>
        <td width = "224"><a href = ""><span th:text = "#{book.buyBook}"></span></a></td>
        <td width = "304"><a href = ""><span th:text = "#{book.cart}"></span></a></td>
        <td width = "304"><a href = ""><span th:text = "#{book.order}"></span></a></td>
        <td width = "304"><span th:text = "#{book.date} + ':' + ${dateString}"/></td>
    </tr>
</table>
```

代码中使用 Thymeleaf 模块的 #{} 消息表达式设置了国际化展示的一部分信息。当对购物车进行国际化设置时，需要在 <span> 标签中设置国际化信息。当对首页进行国际化设置时，这里使用了行内表达式 [[#{book.firstPage}]] 动态获取国际化文件中的信息。

此时国际化配置完成了一半，需要根据浏览器的请求头 Accept-Language 的语言进行页面的回显，但与程序无关，如图 3-24 所示。

图 3-24　Accept-Language 的语言为中文

在浏览器设置中切换需要显示的语言为英语置顶，重新刷新页面就可以显示出英文的信息，如图 3-25 所示。

图 3-25　Accept-Language 的语言为英文

（6）实现国际化的方式有两种，一种是根据浏览器的语言变化；另一种是单击链接来改变语言。接下来需要手动对页面语言进行自定义切换。修改 foot.html 页面，增加中英文切换选项，代码如下所示。

```
<div align="center" th:fragment="foot">
<table width="700" border="0" cellspacing="2" cellpadding="2">
    <tr>
        <td align="center"><span th:text="#{book.development}"></span><br>
            CopyRight@2004<br>
            Email:<a href="mailto:123@163.net">123@163.net</a></td>
    </tr>
    <tr>
        <td align="center">
            <a th:href="@{/login(l='zh_CN')}">中文</a>   
            <a th:href="@{/login(l='en_us')}">English</a>
        </td>
    </tr>
</table>
</div>
```

LocaleResolver（获取区域信息对象）解析器位于国际化的 Locale（区域信息对象），在 Web MVC 自动配置文件中可以看到 Spring Boot 的默认配置。

```
@Bean
@ConditionalOnMissingBean
@ConditionalOnProperty(prefix = "spring.mvc", name = "locale")
public LocaleResolver localeResolver() {
    //容器中没有就自己配置,有的话就使用已有配置
    if (this.mvcProperties.getLocaleResolver() == WebMvcProperties.LocaleResolver.FIXED)
    {    return new FixedLocaleResolver(this.mvcProperties.getLocale());    }
    //接收头国际化分解
    AcceptHeaderLocaleResolver localeResolver = new AcceptHeaderLocaleResolver();
    localeResolver.setDefaultLocale(this.mvcProperties.getLocale());
    return localeResolver;
}
```

AcceptHeaderLocaleResolver 类中获取 Locale 的方法如下：

```java
public Locale resolveLocale(HttpServletRequest request) {
//默认为根据请求头带来的区域信息获取 Locale 进行国际化
    Locale defaultLocale = this.getDefaultLocale();
    if (defaultLocale != null && request.getHeader("Accept-Language") == null) {
        return defaultLocale;
    } else {
        Locale requestLocale = request.getLocale();
        List<Locale> supportedLocales = this.getSupportedLocales();
        if (!supportedLocales.isEmpty() && !supportedLocales.contains(requestLocale)) {
            Locale supportedLocale = this.findSupportedLocale(request, supportedLocales);
            if (supportedLocale != null) {
                return supportedLocale;
            } else {
                return defaultLocale != null ? defaultLocale : requestLocale;
            }
        } else {
            return requestLocale;
        }
    }
}
```

如果想单击链接让国际化资源生效，就需要让自己的 Locale 生效。此时在 it.com.boot.springboot03_05.config 包下创建一个用于定制国际化功能区域信息解析器的自定义配置类 MyLocaleResolver，通过链接获取携带区域的信息。

```java
public class MyLocaleResolver implements LocaleResolver {
//解析请求
    @Override
    public Locale resolveLocale(HttpServletRequest request) {
        String language = request.getParameter("l");       //获取 header 页面传过来的参数
    String head = request.getHeader("Accept-Language");
//Accept-Language: zh-CN,zh;q=0.9,en-US;q=0.8,en;q=0.7
        Locale locale;
        if (!StringUtils.isEmpty(language)) {              //如果请求链接不为空
            //由于格式为 en_us,所以用"_"分割请求参数
    String[] split = language.split("_");
      //截取后把新的属性(国家,地区)封装给 locale
    locale = new Locale(split[0], split[1]);
        }
        else {
            String[] splits = head.split(",");
            String[] split = splits[0].split("-");
            locale = new Locale(split[0], split[1]);
        }
    return locale;
    }
    @Override
    public void setLocale (HttpServletRequest httpServletRequest, HttpServletResponse httpServletResponse, Locale locale) {
    }
}
```

（7）为了让区域化信息能够生效，需要再配置一下这个组件。修改 MyLocaleResolver

类为配置类，并在该类下重新注册一个类型为 LocaleResolver 的 Bean 组件，这样就可以覆盖默认的 LocaleResolver 组件。

```
@Configuration
public class MyLocaleResolver implements LocaleResolver {
    //省略解析请求
    @Bean
    public LocaleResolver localeResolver()
    {
        return new MyLocaleResolver();
    }
}
```

（8）重启服务器，发现单击按钮也可以实现成功切换，中英文页面分别如图 3-26 和图 3-27 所示。

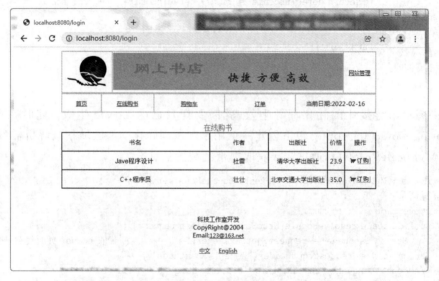

图 3-26　中文页面

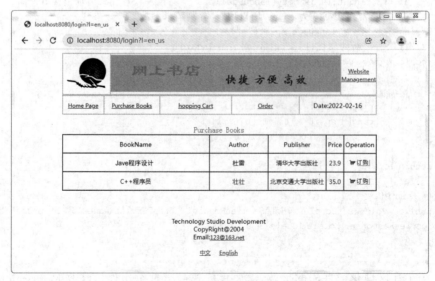

图 3-27　英文页面

3.6　Spring Boot 集成 Spring MVC

如果想在 Spring Boot 中自己定义一些 Handler、Interceptor、ViewResolver 和 MessageConverter 等,该如何完成呢?在 Spring Boot 1.5 版本中,依靠重写 WebMvcConfigurerAdapter 的方法来添加自定义拦截器和消息转换器等。在 SpringBoot 2.0 版本后,该类被标记为 @Deprecated,因此只能靠 WebMvcConfigurer 接口来实现。

WebMvcConfigurer 是一个接口,提供很多自定义的拦截器,如跨域设置、类型转换器等。此接口为开发者预想了很多拦截层面的需求,方便开发者自由选择使用。

Spring Boot 推荐使用 WebMvcConfigurer 接口的实现类来实现代码配置,具体代码如下。

```
@Configuration
public class MyConfigurer implements WebMvcConfigurer {}
```

3.6.1　配置自定义拦截器 Interceptor

Java 里的拦截器是动态拦截 Action 调用的对象,它提供了一种机制,可以使开发者在一个 Action 执行的前后执行一段代码,也可以在一个 Action 执行前阻止其执行。此外,拦截器还提供了一种可以提取 Action 中可重用部分代码的方式。在 Spring AOP 中,拦截器用于在某个方法或字段被访问前进行拦截,然后在其前后加入某些操作。

下面使用 WebMvcConfigurer 接口中的 addInterceptors()方法注册自定义拦截器。

(1) 以项目 springboot0301 和 springboot0305 为基础,创建项目 springboot0306。

(2) 在 config 包下创建一个自定义拦截器类 MyInterceptor,并编写拦截业务代码。实现拦截器需要一个 HandlerInterceptor 接口的实现类,并重写其中的 3 个方法,具体代码如下。

```
@Component
public class MyInterceptor implements HandlerInterceptor {
    //在请求处理前进行调用(Controller 方法调用前)
    @Override
    public boolean preHandle (HttpServletRequest request, HttpServletResponse response, Object handler) throws Exception {
        Object user = request.getSession().getAttribute("user");
        if (user == null) {
            response.sendRedirect("/index");
            return false;
        }
        return true;
    }
    //请求处理后进行调用(Controller 方法调用后),但需要在视图被渲染前
    @Override
    public void postHandle(HttpServletRequest request, HttpServletResponse response, Object handler, ModelAndView modelAndView) throws Exception {
    }
    //在整个请求结束后被调用,即在 DispatcherServlet 渲染了对应的视图后执行(主要是用于进行资源清理工作)
```

```java
    @Override
    public void afterCompletion(HttpServletRequest request, HttpServletResponse response,
Object handler, Exception ex) throws Exception {
    }
}
```

只有拦截器的 preHandle 方法返回 true，postHandle、afterCompletion 才有可能被执行。如果 preHandle 方法返回 false，则该拦截器的 postHandle、afterCompletion 必然不会被执行。

（3）在实现拦截器后，还需要将拦截器注册到 Spring 容器中。在 config 包下创建一个实现实现 WebMvcConfigurer 接口的配置类 MyConfigurer，覆盖其 addInterceptors (InterceptorRegistry registry) 方法。将 Bean 注册到 Spring 容器中，可以选择@Component 或@Configuration，具体代码如下。

```java
@Configuration
public class MyConfigurer implements WebMvcConfigurer {
    @Autowired
    private MyInterceptor myInterceptor;
    @Override
    public void addInterceptors(InterceptorRegistry registry) {
        registry.addInterceptor(myInterceptor)
                .addPathPatterns("/**")                 //所有路径都被拦截
                .excludePathPatterns("/main")           //main 请求不被拦截
                .excludePathPatterns("/index")
                .excludePathPatterns("/**/*.js")        //JS 静态资源不被拦截
                .excludePathPatterns("/**/*.css");      //CSS 静态资源不被拦截
    }
}
```

其中，addPathPatterns 用于设置拦截器的过滤路径规则，excludePathPatterns 用于设置不需要拦截的过滤规则。

（4）修改 BookController，具体代码如下。

```java
@Controller
public class BookController {
    @RequestMapping("/login")
    public String findAll(Model model)
    {
        Calendar calendar = Calendar.getInstance();
        SimpleDateFormat dateFormat = new SimpleDateFormat("yyyy-MM-dd");
        BookDAO bookDao = new BookDAO();
        List bookList = bookDao.queryAllBook();
        model.addAttribute("dateString", dateFormat.format(calendar.getTime()));
        model.addAttribute("bookList", bookList);
        return "booklist";
    }
    @RequestMapping("/index")
    public String toIndex() {
        return "index";
    }
    @RequestMapping("/main")
    public String toMain(HttpSession session) {
        session.setAttribute("user", "userLoginSuccess");
```

```
        return "redirect:login";
    }
}
```

(5) 运行程序。当访问 http://localhost:8080/login 路径时,系统会自动跳转到用户登录页面,这就说明此处定制的自定义拦截器生效了。由于对 CSS 和 JS 进行了不拦截设置,所以在登录页面中可以进行用户名和密码的验证。只有用户信息填入完整,单击"登录"按钮,才能回到图书首页。

3.6.2 跳转指定页面

用传统方式写 Spring MVC 时,如果需要访问一个页面,必须要写 Controller 类,然后再写一个方法跳转到页面,这样会很麻烦。例如,"/index"可以通过重写 WebMvcConfigurer 中的 addViewControllers 方法达到效果。

addViewControllers 方法可以实现将一个无业务逻辑的请求直接映射为视图,不需要编写控制器,从而简化了页面跳转。

修改 3.6.1 节的 MyConfigurer 配置类,在其实现类中重写 addViewControllers 方法,具体代码如下:

```
@Configuration
public class MyConfigurer implements WebMvcConfigurer {
    @Override
    public void addViewControllers(ViewControllerRegistry registry) {
        registry.addViewController("/index").setViewName("index");
        registry.addViewController("/index.html").setViewName("index");
    }
}
```

此处重写 addViewControllers 并不会覆盖 WebMvcAutoConfiguration 中的 addViewControllers(在此方法中,Spring Boot 将"/"映射至 index.html),这意味着自己的配置和 Spring Boot 的自动配置同时有效,这也是推荐该配置方式的原因。

3.7 Spring Boot 处理 JSON 数据

视频讲解

在后台的开发过程中,不可避免地会遇到一系列对 JSON 数据的返回,此时需要提供各种各样的数据。一般情况下数据类型最常用的是 JSON 和 XML,这里主要讲解 Spring Boot 怎样进行 JSON 数据的返回及对一些特殊情况的处理。

JSON 是目前主流的前后端数据传输方式,在 Spring MVC 中使用消息转换器 HttpMessageConverter 对 JSON 的转换提供了很好的支持,在 Spring Boot 中对相关配置做了进一步简化。

```
<dependency>
    <groupId>org.springframework.boot</groupId>
    <artifactId>spring-boot-starter-web</artifactId>
</dependency>
```

上述依赖中默认加入了 Jackson-databind 作为 JSON 处理器,此时不需要添加额外的 JSON 处理器就可以返回 JSON。这是 Spring Boot 自带的处理方式,如果采用这种方式,对于字段忽略、日期格式化等都可以使用注解实现。

```
public class Book {
    @JsonIgnore                                     //过滤该属性
    protected Float price;
    @JsonFormat(pattern = "yyyy-MM-dd")             //格式化输出该属性
    private Date publicationDate;
}
```

Spring Boot 在处理 JSON 数据时,需要用到两个重要的 JSON 格式转换注解,分别是 @ResponseBody 和 @RequestBody。

(1) @ResponseBody 注解的作用:将 controller 的方法返回的对象通过适当的转换器转换为指定的格式,并将其写入 response 对象的 body 区,通常用来返回 JSON 数据或 XML 数据。

(2) @RequestBody 注解的作用:在形参列表上,将前台发送的固定格式的数据(XML 数据或 JSON 数据等)封装为对应的 JavaBean 对象,封装时使用系统默认配置的 HttpMessageConverter(消息转换器)进行解析,然后封装到形参上。

常见的 JSON 处理器除了 Jackson-databind 外,还有 Gson 和 FastJson 两种,在使用时需要添加相应的依赖。这里以 FastJson 为例,讲解 JSON 处理器的使用方法。

fastjson.jar 是阿里巴巴开发的一款专门用于 Java 开发的包,可以方便地实现 JSON 对象与 JavaBean 对象的转换、JavaBean 对象与 JSON 字符串的转换,以及 JSON 对象与 JSON 字符串的转换。

下面通过示例讲解 JSON 数据的处理过程。

(1) 以项目 springboot0306 为基础,创建项目 springboot0307。

(2) 打开 pox.xml 文件,去除 jackson-databind 依赖,加入 FastJson 依赖。

```xml
<dependency>
    <groupId>org.springframework.boot</groupId>
    <artifactId>spring-boot-starter-web</artifactId>
    <exclusions>
        <exclusion>
            <groupId>com.fasterxml.jackson.core</groupId>
            <artifactId>jackson-databind</artifactId>
        </exclusion>
    </exclusions>
</dependency>
<dependency>
    <groupId>com.alibaba</groupId>
    <artifactId>fastjson</artifactId>
    <version>1.2.75</version>
</dependency>
```

(3) FastJson 需要自己配置 HttpMessageConverter,打开 MyConfigurer 配置类,在其实现类中,重写 configureMessageConverters 方法,具体代码如下。

```java
@Override
public void configureMessageConverters(List<HttpMessageConverter<?>> converters) {
    //创建 FastJson 的消息转换器
    FastJsonHttpMessageConverter convert = new FastJsonHttpMessageConverter();
```

```java
//创建 FastJson 的配置对象
FastJsonConfig config = new FastJsonConfig();
//对 JSON 数据进行格式化
formatting(SerializerFeature.PrettyFormat);
convert.setFastJsonConfig(config);
converters.add(convert);
}
```

(4) 修改"/login"请求,使其返回 JSON 格式的数据。

```java
@RequestMapping("/login")
@ResponseBody
public String findAll( )
{
    Calendar calendar = Calendar.getInstance();
    SimpleDateFormat dateFormat = new SimpleDateFormat("yyyy-MM-dd");
    BookDAO bookDao = new BookDAO();
    List bookList = bookDao.queryAllBook();
    JSONObject jsonObject = new JSONObject();
    jsonObject.put("dateString", dateFormat.format(calendar.getTime()));
    jsonObject.put("booklist",bookList );
    return jsonObject.toJSONString();
}
```

(5) 修改 booklist.html 页面,使其在页面中显示 AJAX 请求到的数据。

```html
<script type="application/javascript">
    $(function () {
       var tbody = document.getElementById("tbody-result");
       var date = document.getElementById("date");
       $.ajax({
                url: "login",
                type: "get",
//如果当前为 JSON 格式,需要修改此属性,定义发送请求的数据格式为 JSON 字符串
                contentType: "application/json; charset=UTF-8",
                //定义回调响应的数据格式为 JSON 字符串,此属性可以省略
                dataType: "json",
                //成功响应的结果
                success: function (data) {
                    var inf = "";
                    var bookList = data.booklist;
                    for (var i = 0; i < bookList.length; i++) {
                        var book = bookList[i];
                        inf += "<tr><td height='32'><div align='center'>" + book.title +
                            "</div></td><td><div align='center'>" + book.author +
                            "</div></td><td><div align='center'>" + book.publisher +
                            "</div></td><td><div align='center'>" + book.price +
                            "</div></td><td><div align='center'><a href=''><img src='images/buy.gif' width='45' height='16' border='0'></a></div></td></tr>"
                    }
                    tbody.innerHTML = inf;
                    date.innerText += data.dateString;
                }
        };
```

```
            error: function () {
                alert(("操作失败"));
            }
        }
    );
})
</script>

<table width = "700" border = "1" cellpadding = "0" cellspacing = "0">
    <tr>省略标题行</tr>
    <tbody id = "tbody-result"></tbody>
</table>
```

(6) 运行程序,效果图如图 3-26 所示。

3.8 Spring Boot 实现 RESTful 风格的 Web 应用

视频讲解

RESTful 架构风格是目前最流行的一种架构风格,它结构清晰、符合标准、易于理解、扩展方便,所以在 Web 开发中经常被使用。REST 的全称是 Representational State Transfer,译作"表现层状态转化"。

RESTful 架构是对 MVC 架构改进后所形成的一种架构,通过使用事先定义好的接口与不同的服务联系起来。在 RESTful 架构中,浏览器使用 POST、DELETE、PUT 和 GET 这 4 种请求方式分别对指定的 URL 资源进行增、删、改、查操作。因此,RESTful 通过 URI 实现对资源的管理和访问,具有扩展性强、结构清晰的特点。

RESTful 架构将服务器分成前端服务器和后端服务器两部分,前端服务器为用户提供无模型的视图,后端服务器为前端服务器提供接口。浏览器向前端服务器请求视图,通过视图中包含的 AJAX 函数发起接口请求并获取模型。

RESTful 是一种对 URL 进行规范的编码风格,通常一个网址对应一个资源,访问形式类似 http://xxx.com/xx/{id}/{id}。

举个例子,在某购物网站上买手机时会有很多品牌选择,而每种品牌下又有很多型号,此时 https://mall.com/mobile/iPhone/6 可以代表 iphone6,而 https://mall.com/mobile/iPhone/7 和 https://mall.com/mobile/iPhone/8 分别代表 iphone7 和 iphone8。

在进行 Web 开发的过程中,method 常用的值是 GET 和 POST。可事实上,method 值还可以是 PUT 和 DELETE 等其他值。既然 method 值如此丰富,那么就可以考虑使用同一个 URL,但是约定不同的 method 来实施不同的业务,这就是 RESTful 的基本考虑。

CRUD 是最常见的操作,在使用 RESTful 风格前,通常的增加做法如下:

/addCategory?name = xxx

可是在使用 RESTful 风格后,增加做法就变为:

/categories

传统风格和 RESTful 风格的对比如表 3-1 所示,URL 使用相同的"/categories"语句,区别只是在于 method 值不同,服务器根据 method 的不同来判断浏览器期望进行的业务行为。

表 3-1 传统风格和 RESTful 风格的对比

功能	传统风格		RESTful 风格	
	URL	method	URL	method
增加	/addCategory?name=xxx	POST	/categories	POST
删除	/deleteCategory?id=123	GET	/categories/123	DELETE
修改	/updateCategory?id=123&name=yyy	POST	/categories/123	PUT
获取	/getCategory?id=123	GET	/categories/123	GET
查询	/listCategory	GET	/categories	GET

为了实现 RESTful API 接口，Spring Boot 提供了以下注解，对请求参数和返回数据格式进行封装，方便用户快速开发。

(1) @RestController 一般用于 Controller 类上，指定所有接口返回的数据都是 text/json 格式。

(2) @ResponseBody 用于方法上，指定接口返回 text/json 格式，如果使用 @RestController 就没有必要使用@ResponseBody。

(3) @GetMapping 用于方法上，是一个组合注解，与 @RequestMapping(method=RequestMethod.GET)作用一致。

(4) @PostMapping 用于方法上，是一个组合注解，与 @RequestMapping(method=RequestMethod.POST)作用一致。

(5) @PutMapping 用于方法上，是一个组合注解，与 @RequestMapping(method=RequestMethod.PUT)作用一致。

(6) @DeleteMapping 用于方法上，是一个组合注解，与 @RequestMapping(method=RequestMethod.DELETE)作用一致。

(7) @RequestParam 用于方法上，映射请求参数到 Java 方法的参数，当前端传递的参数与后台自定义的参数不一致时，可以使用 name 属性来标记。

(8) @PathVariable 用于方法上，映射 URL 片段到 Java 方法的参数。

下面通过一个例题来讲解 Spring Boot 如何实现 RESTful。

(1) 以项目 springboot0307 为基础，创建项目 springboot0308。

(2) 创建 Controller 类。

```
@RestController
public class UserController {
    @GetMapping("/user/{username}")
    public String toMain(@PathVariable String username, String password,HttpSession session ) {
        JSONObject jsonObject = new JSONObject();
        if (username.equals("admin")&&password.equals("123456")) {
            session.setAttribute("user", "userLoginSuccess");
            jsonObject.put("flag", true);
        }
        else
            jsonObject.put("flag", false);
        return jsonObject.toJSONString();
    }
    @DeleteMapping("/user/{username}")
    public String deleteUser( @PathVariable String username ,HttpSession session ) {
```

```java
        JSONObject jsonObject = new JSONObject();
        System.out.println(username);
        jsonObject.put("flag", "删除成功");
        return jsonObject.toJSONString();
    }
    @PostMapping("/user")
    public String addUser( @RequestBody String json ) {
        JSONObject jsonObject = JSONObject.parseObject(json);
        String username = jsonObject.getString("username");
        String password = jsonObject.getString("password");
        System.out.println(username);
        System.out.println(password);
        jsonObject.put("flag", "增加成功");
        return jsonObject.toJSONString();
    }
    @PutMapping("/user")
    public String updateUser(User user) {
        JSONObject jsonObject = new JSONObject();
        System.out.println(user);
        jsonObject.put("flag", "修改成功");
        return jsonObject.toJSONString();
    }
}
```

(3)编写相应的 AJAX 提交,这里只列出查询和添加两类,修改和删除类似,具体代码如下。

```javascript
function checkform() {
    var username = document.getElementById("userid").value;
    var password = document.getElementById("password").value;
    var json = {"password": password};
    $.ajax({
        url: "user/" + username,
        type: "get",
        data: json,
        dataType: "json",
        //成功响应的结果
        success: function (data) {
            if (data.flag) {
                alert("登录成功");
                window.location = "tologin";
            }
            else
                alert("登录失败")
        },
        error: function () {
            alert(("操作失败"));
        }
    }
    );
}
function addform() {
    var username = $("#userid").val();
    var password = $("#password").val() ;
    alert(username)
    var json = {"username":username,"password": password};
```

```
$.ajax({
    url: "user",
    type: "post",
    data: JSON.stringify(json),
    contentType: "application/json; charset = UTF - 8",
    dataType: "json",
    //成功响应的结果
    success: function (data) {
        if (data.flag) {
            alert("添加成功");
        }
        else
            alert("添加失败")
    },
    error: function () {
        alert(("操作失败"));
    }
});
}
```

(4) 运行程序,效果图如图 3-28～图 3-31 所示。

图 3-28　登录成功效果图

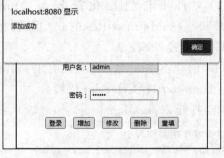

图 3-29　添加成功效果图

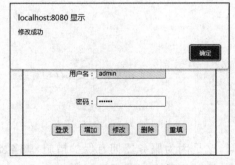

图 3-30　修改成功效果图

图 3-31　删除成功效果图

3.9　Spring Boot 文件上传和下载

在 Web 应用中,对多媒体文件的操作非常常见,文件的上传与下载尤为如此。本节将通过项目的新增图片和附件的上传、下载模块,带领大家熟悉该类型功能的开发流程。

3.9.1 文件上传

Spring Boot 通常为服务提供者,但是在一些场景中还是要用到文件上传下载这种"非常规"操作。如何在 Spring Boot 中实现文件的上传与下载功能呢?回顾一下在 Spring MVC 中的操作,需要在 Spring MVC 的配置文件中增加文件上传的 Bean 的配置。

```
<bean id="multipartResolver"
    class="org.springframework.web.multipart.commons.CommonsMultipartResolver"/>
```

在 Spring MVC 中完成 Bean 配置后,在后台对应的处理方法中就可以直接获取文件的输入流。而对于 Spring Boot 来说,不需要配置文件上传的解析类,因为 Spring Boot 已经注册好了。

Java 中的文件上传涉及两个组件,一个是 CommonsMultipartResolver,另一个是 StandardServletMultipartResolver。CommonsMultipartResolver 使用 commons-fileupload 来处理 multipart 请求,而 StandardServletMultipartResolver 则基于 Servlet 3.0 来处理 multipart 请求。因此,若使用 StandardServletMultipartResolver 处理请求,则不需要添加额外的 JAR 包。Tomcat 7.0 开始支持 Servlet 3.0,而 Spring Boot 2.6.3 内嵌的 Tomcat 为 Tomcat 9.0.56,因此可以直接使用 StandardServletMultipartResolver。在 Spring Boot 提供的文件上传自动化配置类 MultipartAutoConfiguration 中,默认也是采用 StandardServletMultipartResolver。因此,在 Spring Boot 中上传文件甚至可以做到零配置,具体上传过程如下。

(1)创建 Spring Boot Web 应用 springboot0309,并添加 spring-boot-starter-web 和 spring-boot-starter-thymeleaf 依赖。

(2)打开 application.properties 文件,添加如下配置,完成对上传文件大小的限制。

```
#设置单个文件大小
spring.servlet.multipart.max-file-size=5MB
#设置单次请求文件的总大小
spring.servlet.multipart.max-request-size=500MB
```

(3)创建 User 类。

```
public class User {
    private String username,password;
    private String tupian;                              //上传头像需要保存相对地址
    //省略 get 和 set 方法
}
```

(4)创建 MyConfig 配置类。

图片上传无法立即显示而需要重启才能访问,这是因为服务器拥有保护措施,服务器不能对外部暴露真实的资源路径。为了解决该问题,需要配置虚拟路径映射进行访问。

```
@Configuration
public class MyConfig implements WebMvcConfigurer {
    @Override
    public void addResourceHandlers(ResourceHandlerRegistry registry) {
        //获取文件的真实路径
        String path = System.getProperty("user.dir") + "\\src\\main\\resources\\static\\upload\\";
```

```
    ///upload/** 是对应 resource 下的工程目录
    registry.addResourceHandler("/upload/**").addResourceLocations("file:" + path);
  }
}
```

其中 System.getProperty("user.dir")是当前项目路径,如图 3-32 所示。

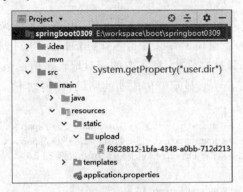

图 3-32 当前项目路径

(5) 创建上传页面。

在 resources 目录下的 static 目录中创建一个 zhuce.html 文件,完成个人信息的注册,代码如下。上传接口是/zhuce,注意请求方法是 POST,enctype 是 multipart/form-data。

```html
<form th:action="@{/zhuce}" method="post" enctype="multipart/form-data">
    <label>姓名</label><input type="text" name="username"><br>
    <label>密码</label><input type="password" name="password"><br>
    <label>图片</label><input type="file" name="file"><br>
    <input type="submit" value="上传">
</form>
```

(6) 创建文件上传处理接口。

```java
@Controller
public class UserController {
  @GetMapping("/zhuce")
  public String zhuce() {
    return "zhuce";
  }
  @PostMapping("/zhuce")
  public String tijiao(User user, MultipartFile file, Model model) {
    if (!file.isEmpty()) {
      String fileName = UUID.randomUUID() + "_" + file.getOriginalFilename();
      String path = System.getProperty("user.dir");
      File filePath = new File(path, "\\src\\main\\resources\\static\\upload");
      if (!filePath.isDirectory()) {
          filePath.mkdirs();
      }
      try {
          File userFile = new File(filePath, fileName);
          file.transferTo(userFile);
          user.setTupian("/upload/" + fileName);
      } catch (IOException e) {
              e.printStackTrace();
```

            }
        }
        model.addAttribute("user", user);
        return "permanager";
    }
}
```

规划上传文件的保存路径为项目运行目录下的 upload 文件夹，如果该文件夹不存在，则新建 upload 文件夹。为了避免上传文件重名，这里通过随时生成 uid 给上传文件名重新命名。将上传文件保存到目标文件中，把头像的相对地址赋值给当初用户的图片属性上。

（7）创建 permanager.html，显示上传成功的结果，代码如下。

```html
<div th:object="${user}">
    <label>姓名</label><span th:text="*{username}"></span><br>
    <label>密码</label><span th:text="*{password}"></span><br>
    <label>图片</label><img th:src="@{*{tupian}}"><br>
</div>
```

（8）运行程序，在浏览器地址栏中输入 http://localhost:8080/zhuce，选择要上传的头像，上传界面如图 3-33 所示。单击"上传"按钮，成功上传头像后显示注册结果，如图 3-34 所示。

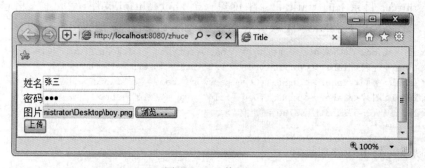

图 3-33　上传界面

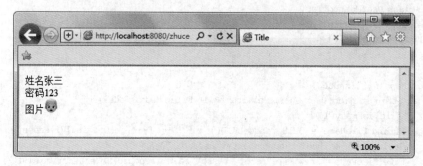

图 3-34　显示上传成功界面

3.9.2　文件下载

在 Spring Boot 项目中，可能会存在让用户下载文档的需求，比如让用户下载 readme 文档来更好地了解该项目的概况或使用方法。因此，需要为用户提供可以下载文件的 API，将用户希望获取的文件作为下载资源返回给前端。
```

下面以项目 springboot0309 为基础，继续分析下载功能。

(1) 配置 application.xml。

```
file.docDir: /src/main/resources/static/upload
```

该路径就是待下载文件存放在服务器上的目录，即相对路径。

(2) 将属性与实体类自动绑定。

Spring Boot 中的注解 @ConfigurationProperties 可以将 application 中定义的属性与 POJO 类自动绑定，为此需要定义一个 POJO 类来进行 application 中 file.docDir 的配置绑定。

```
@ConfigurationProperties(prefix = "file")
@Component
public class FileProperties {
    private String docDir;
    public String getDocDir() {
        return docDir;
    }
    public void setDocDir(String docDir) {
        this.docDir = docDir;
    }
}
```

注解 @ConfigurationProperties(prefix = "file") 在 Spring Boot 应用启动时将以 file 为前缀的属性与 POJO 类绑定，也就是将 application.xml 中的 file.docDir 与 FileProperties 中的字段 docDir 进行绑定。

(3) 编写自定义配置类。

打开 MyConfig 配置类，生成下载文件所在目录的配置类，其返回值对象返回值会作为组件添加到 Spring 容器中。

```
@Autowired
private FileProperties fileProperties;
@Bean
public File getFile() {
  String path = System.getProperty("user.dir");
  File fileDir = new File(path, fileProperties.getDocDir());
  return fileDir;
}
```

(4) 编写控制器。

```
@Controller
public class UserController {
    @Autowired
    private File fileDir;
    @GetMapping("/showdownload")
    public String showDownLoad(Model model) {
        File[] fileList = fileDir.listFiles();
        model.addAttribute("fileList", fileList);
        return "download";
    }
    @GetMapping("/download")
    public void toDownLoad(String filename, HttpServletResponse response) {
```

```
            try {
                //通过流读取文件
                FileInputStream is = new FileInputStream(new File(fileDir, filename));
        //获得响应流
                ServletOutputStream os = response.getOutputStream();
                //设置响应头信息
                response.setHeader("content-disposition", "attachment;fileName=" + URLEncoder.
encode(filename, "UTF-8"));
                //通过响应流将文件输入流读取的文件写出
                IOUtils.copy(is, os);
                //关闭流
                IOUtils.closeQuietly(is);
                IOUtils.closeQuietly(os);
            } catch (Exception ex) {
            }
        }
    }
```

(5) 编写文件下载程序。

在前端 download.html 页面中,可以使用 a 标签来下载文件,但要注意在 a 标签中定义 filename 参数来规定这是下载文件。

```
<table>
    <tr>
        <th>序号</th>
        <th>文件</th>
    </tr>
    <tr th:each="file,fileState: ${fileList}">
        <td><span th:text="${fileState.count}"></span></td>
        <td>
            <a th:href="@{/download(filename = ${file.name})}">
                <span th:with="tempFileName = ${file.name.substring(file.name.lastIndexOf('_') + 1)}">
                <span th:text="${tempFileName}"></span>
            </span>
            </a>
        </td>
    </tr>
</table>
```

(6) 运行程序,在浏览器地址栏中输入 http://localhost:8080/showdownload,进入下载页面,选择所需要下载的文件进行下载,如图 3-35 所示。

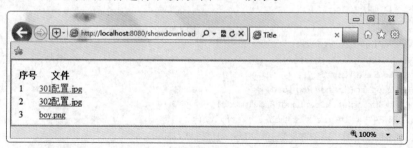

图 3-35 文件下载页面效果

3.10 Spring Boot 的异常统一处理

在程序中出现错误是难以避免的，经验再丰富的程序员编写的程序也会出现错误。在 Java 中，程序出现错误会抛出"不正常信息"（Throwable）。Throwable 又被分为"错误"（Error）和"异常"（Exception）。有别于人为失误造成的"故障"（Bug），异常在程序中代表的是出现了当前代码无法处理的状况。例如，在一个对象不存在（值为 Null）的情况下，调用该对象的某个方法引发了空指针；用户输入了一段 URL，但并没有找到对应的资源；在一段计算过程中，0 被当作除数，等等。完善的错误处理，使程序不会意外崩溃甚至能友好地提示用户进行正确操作，这是让程序变得越发健壮的重要处理步骤。

在 Java 开发中，异常特别是检查型异常（Checked Exception），通常需要进行 try/catch 处理。而在基于 Spring Boot 的开发过程中，异常处理有了更多的处理方式。

3.10.1 自定义 error 页面

使用 Web 应用时，在请求处理过程中发生错误是非常常见的情况。Spring Boot 提供了一个默认的映射/error，在抛出异常后，会转到该请求中处理，并且该请求有一个全局的错误页面来展示异常内容。例如，启动某个项目，在浏览器中随便输入一个访问地址，由于地址不存在，Spring Boot 会跳转到错误页面，如图 3-36 所示。

Whitelabel Error Page

This application has no explicit mapping for /error, so you are seeing this as a fallback.

Sat Jan 29 16:27:50 CST 2022
There was an unexpected error (type=Not Found, status=404).
No message available

图 3-36　错误页面

虽然 Spring Boot 提供了默认的错误页面映射，但是在实际应用中，图 3-36 所示的错误页面对用户来说并不友好，需要自己实现异常提示。接下来将演示自己如何实现错误提示页面。

(1) 创建 Spring Boot 项目 springboot03_10，并添加相应的依赖。

(2) 创建控制器类，该类即为了演示错误而定制。

```
@Controller
public class TestController {
  @GetMapping("/index")
  public String toIndex()
  {
    int result = 1/0;
    return "index";
  }
}
```

(3) 定制错误页面。

在 Spring Boot 中定制错误页面有 3 种方法，分别如下。

① 使用精确匹配的方式，将错误页面命名为"错误状态码.html"，并放在模板引擎

templates/error 文件夹下。当发生访问错误时，会跳转到对应状态码的页面，如 404.html 和 500.html。也可以使用模糊匹配的方式来定义错误页面的名称，将错误页面命名为 4xx.html 和 5xx.html，以此匹配对应类型的所有错误。精确匹配的查找方式要优先于模糊匹配的方式。

② 如果 templates/error 目录下没有自定义的错误页面，那么需要在 static/error 目录下定义 4xx.html 或 5xx.html 页面。

③ 直接在 templates 目录下创建 error.html 页面，这样当访问错误或异常时，可以自动将该页面作为错误页面。

Spring Boot 为错误页面提供了以下属性。

- timestamp：时间戳。
- status：状态码。
- error：错误提示。
- exception：异常对象。
- message：异常消息。
- errors：JSR303 数据校验的错误。

在 5xx.html 页面中编写如下代码。

```html
<body background="500.jpeg">
<div>
    <p><b>错误发生时间：</b>
    <span th:text="${#dates.format(timestamp,'yyyy-MM-dd')}"></span></p>
    <p><b>错误状态码：</b><span th:text="${status}"></span></p>
    <p><b>异常消息：</b>[[ ${message} ]]</p>
    <p><b>错误提示：</b>[[ ${error} ]]</p>
</div>
</body>
```

（4）运行程序。

在成功运行项目后，访问 http://localhost:8080/hello。由于服务器找不到请求的网页，Spring Boot 便会找到 src/main/resources/templates/error 目录中的 404.html 页面，运行效果如图 3-37 所示。

图 3-37　404 错误页面

继续访问 http://localhost:8080/index。该请求中计算除法发生了异常，而该方法仅抛出了 exception 异常，并没有处理异常。当 Spring Boot 发现有异常抛出且没有处理时，将在 src/main/resources/templates/error 目录下找到 500.html 页面并显示异常信息，运行效果如图 3-38 所示。

图 3-38　500 错误页面

从上述运行结果可以看出，使用自定义错误页面并没有真正处理异常，只是将异常或错误信息显示给客户端，因为在服务器控制台上同样抛出了异常，如图 3-39 所示。

```
java.lang.ArithmeticException: / by zero
    at it.com.boot.springboot03_10.TestController.toIndex(TestController.java:11)
    at javax.servlet.http.HttpServlet.service(HttpServlet.java:655) ~[tomcat-embed
```

图 3-39　异常信息

3.10.2　@ExceptionHandler 注解

不难发现，使用自定义 error 页面并没有真正处理异常，在本节将@ExceptionHandler 注解处理异常。该注解主要用于在 Controller 层面进行相同类型的异常处理，在对应 Controller 类中定义异常处理方法，并为其使用@ExceptionHandler 注解。Spring 会检测到该注解，并将该方法注册为对应异常类及其子类的异常处理程序。异常处理的示例代码如下。

```
@ExceptionHandler()
public String handleException2(Exception ex) {
    System.out.println("抛出异常:" + ex);
    ex.printStackTrace();
    String resultStr = "异常：默认";
    return resultStr;
}
```

使用该注解的方法可以拥有非常灵活的签名，包括以下类型：

- 异常类型（Throwable）：可以选择一个大概的异常类型。例如，示例里的签名可以改为"Throwable e"或"Exception e"，也可以改为一个具体的异常类型。
- 请求与响应对象（Request/Response）：比如 ServletRequest 和 HttpServletRequest。
- InputStream/Reader：用于访问请求的内容。
- OutputStream/Writer：用户访问响应的内容。

- Model：作为从该方法返回 Model 的替代方案。

在@ExceptionHandler 注解中可以添加参数，参数是某个异常类的 class，代表这个方法专门处理该类异常，示例代码如下：

```
@ExceptionHandler(NumberFormatException.class)
public String handleException(Exception ex) {
    System.out.println("抛出异常:" + ex);
    ex.printStackTrace();
    String resultStr = "异常：NumberFormatException";
    return resultStr;
}
```

此时注解的参数是 NumberFormatException.class，表示只有方法抛出 NumberFormatException 时，才会调用该方法。

当异常发生时，Spring 会选择最接近抛出异常的处理方法。

例如，NumberFormatException 异常，该异常有父类 RuntimeException 和 Exception，如果分别定义异常处理方法，@ExceptionHandler 将分别使用这 3 个异常作为参数，示例代码如下：

```
@ExceptionHandler(NumberFormatException.class)
public String handleException(Exception ex) {
    System.out.println("抛出异常:" + ex);
    ex.printStackTrace();
    String resultStr = "异常：NumberFormatException";
    return resultStr;
}

@ExceptionHandler()
public String handleException2(Exception ex) {
    System.out.println("抛出异常:" + ex);
    ex.printStackTrace();
    String resultStr = "异常：默认";
    return resultStr;
}

@ExceptionHandler(RuntimeException.class)
public String handleException3(Exception ex) {
    System.out.println("抛出异常:" + ex);
    ex.printStackTrace();
    String resultStr = "异常：RuntimeException";
    return resultStr;
}
```

当代码抛出 NumberFormatException 时，调用的方法将是注解参数 NumberFormatException.class 的方法，即 handleException()；而当代码抛出 IndexOutOfBoundsException 时，调用的方法将是注解参数 RuntimeException 的方法，即 handleException3()。

标识了@ExceptionHandler 注解的方法，其返回值类型与标识了@RequestMapping 注解的方法是相同的，可参见 @RequestMapping 的说明。例如，默认返回 Spring 的 ModelAndView 对象，也可以返回 String，这时的 String 是 ModelAndView 的路径，而不是字符串本身。

有些情况下会给标识了@RequestMapping 的方法添加@ResponseBody，比如使用 AJAX 的场景会直接返回字符串。异常处理类也可以如此操作，在添加@ResponseBody 注解后，可以直接返回字符串，示例代码如下：

```
@ExceptionHandler(NumberFormatException.class)
@ResponseBody
public String handleException(Exception ex) {
    System.out.println("抛出异常:" + ex);
    ex.printStackTrace();
    String resultStr = "异常: NumberFormatException";
    return resultStr;
}
```

一个 Spring Boot 应用中往往存在多个控制器，不适合在每个控制器中添加使用@ExceptionHandler 注解修饰的方法进行异常处理。传统的做法是定义一个控制器父类（如 BaseController），它包含了执行共同操作的方法，其他的控制器类（如 ControllerA 和 ControllerB）继承这个控制器父类。图 3-40 显示了控制器父类和控制器子类的关系。

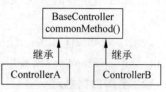

图 3-40 控制器父类和控制器子类的关系

3.10.3 @ControllerAdvice 注解

继承是提高控制器类的代码可重用性的有效手段，但是它有一个缺陷，那就是由于 Java 语言不支持多继承，当控制器类继承了一个控制器父类后，就不能再继承其他的类。Spring MVC 框架提供了另一种方式来为多个控制器类提供共同的方法：利用@ControllerAdvice 注解定义一个控制器增强类。

控制器增强类并不是控制器类的父类。在程序运行时，Spring MVC 框架会把控制器增强类的方法代码块动态注入其他控制器类中，通过这种方式增强控制器类的功能。图 3-41 显示了控制器增强类（如 MyControllerAdvice）和控制器类的关系。

图 3-41 控制器增强类和控制器类的关系

@ControllerAdvice 注解是 Spring 3.2 中新增的注解，学名是 Controller 增强器，作用是给 Controller 控制器添加统一的操作或处理。

对于@ControllerAdvice，比较熟知的用法是结合@ExceptionHandler 用于全局异常的处理，但其作用不止于此。ControllerAdvice 拆开来就是 Controller Advice，Advice 在 Spring 的 AOP 中是用来封装一个切面所有属性的，包括切入点和需要织入的切面逻辑。这里 ControllerAdvice 也可以这么理解，其抽象级别应该是用于对 Controller 进行切面环绕的，而具体的业务织入方式则是通过结合其他的注解来实现的。@ControllerAdvice 是在类上声明的注解，其用法主要有 3 点，灵活使用这 3 个功能可以简化很多工作。需要注意的是，这是 Spring MVC 提供的功能，在 Spring Boot 中可以直接使用。下面介绍@ControllerAdvice 的具体用法。

1. 全局异常处理

使用@ControllerAdvice 实现全局异常处理，只需要定义类并添加该注解即可。定义类的方式如下：

```java
@ControllerAdvice
public class MyGlobalExceptionHandler {
    @ExceptionHandler(Exception.class)
    public ModelAndView customException(Exception e) {
        ModelAndView mv = new ModelAndView();
        mv.addObject("message", e.getMessage());
        mv.setViewName("myerror");
        return mv;
    }
}
```

在该类中,可以定义多个方法,不同的方法处理不同的异常,如专门处理空指针的方法、专门处理数组越界的方法等,也可以直接像上面代码一样,在一个方法中处理所有的异常信息。

@ExceptionHandler 注解用来指明异常的处理类型,即如果这里指定为 NullpointerException,则数组越界异常就不会进入该方法。

2. 全局数据绑定

全局数据绑定功能可以用来做一些初始化的数据操作,可以将一些公共的数据定义在添加了 @ControllerAdvice 注解的类中。这样,在每一个 Controller 的接口中,就都能够访问这些数据。

定义全局数据如下:

```java
@ControllerAdvice
public class MyControllerAdvice {
    @ModelAttribute(name = "colors")
    public Map<String,String> setColors() {
        HashMap<String,String> colors = new HashMap<String,String>();
        colors.put("RED", "红色");
        colors.put("BLUE", "蓝色");
        colors.put("GREEN", "绿色");
        return colors;
    }
}
```

当程序运行时,Spring MVC 框架会把 MyControllerAdvice 类的 setColors() 方法动态注入其他控制器类中,因此其他控制器类自动拥有了该方法。例如,在 TestAttributeController 类中可以直接访问 Model 中的 colors 属性。

```java
@RequestMapping(value = "/testColor")
public String testColor(@ModelAttribute("colors") Map<String,String> colors, @ModelAttribute("userName") String name){
    System.out.println(name + "'s favourite color:" + colors.get("RED"));
    return "result";
}
```

通过浏览器访问 http://localhost:8080/helloapp/testColor?name=Tom,testColor() 方法会在服务器端打印"TOM's favourite color:红色"。

对控制器添加通知有以下几种方式:

(1) 只对一部分控制器添加通知,如某个包下的控制器。

```java
@ControllerAdvice(basePackages = {"com.example"})
//@ControllerAdvice(com.example)
```

```
public class MyControllerAdvice2{ … }
```

basePackages：指定一个或多个包，这些包及其子包下的所有 Controller 都被该 @ControllerAdvice 管理。

（2）不固定包名，只想把包里的某个类传进去。

```
@ControllerAdvice(basePackageClasses = MainController.class)
public class MyControllerAdvice2{ … }
```

basePackageClasses：basePackages 的一种变形，指定一个或多个 Controller 类，这些类所属的包及其子包下的所有 Controller 都被该 @ControllerAdvice 管理。

（3）只对某几个控制器添加通知。

```
@ControllerAdvice(assignableTypes = {PersonController.class,MainController.class})
public class MyControllerAdvice2{ … }
```

3. 全局数据预处理

考虑有两个实体类为 Car 和 Driver，分别定义如下：

```
public class Car {
    private String name;
    private String type;
    //省略 get 和 set 方法
}
public class Driver {
    private String name;
    private Integer age;
    //省略 get 和 set 方法
}
```

定义一个数据添加接口如下：

```
@PostMapping("/car")
public void addCar(Car car , Driver driver ) {
    System.out.println(car);
    System.out.println(driver);
}
```

此时，添加操作会产生问题，因为两个实体类都有一个 name 属性，从前端传递时无法区分。通过 @ControllerAdvice 的全局数据预处理可以解决该问题，解决步骤如下：

（1）给接口中的变量取别名。

```
@RestController
public class TestController {
  @GetMapping("/index")
  public void index(@ModelAttribute("c") Car car, @ModelAttribute("d") Driver driver) {
    System.out.println(car);
    System.out.println(driver);
    }
}
```

（2）进行请求数据预处理。

在 @ControllerAdvice 标记的类中添加如下代码：

```
@ControllerAdvice(value = "com.example.controller")
public class TestControllerAdvice {
```

```
        @InitBinder("c")
        public void b(WebDataBinder binder) {
            binder.setFieldDefaultPrefix("c.");
        }
        @InitBinder("d")
        public void a(WebDataBinder binder) {
            binder.setFieldDefaultPrefix("d.");
        }
    }
```

@InitBinder("c")注解表示该方法用来处理与 Car 相关的参数,在方法中给参数添加一个 c 前缀,即请求参数要有 c 前缀。

(3)发送请求。

请求发送时,通过给不同对象的参数添加不同的前缀,可以实现参数的区分。

请求地址 http://127.0.0.1:8080/index?c.name=保时捷&c.type=赛跑&d.name=全爷&d.age=35
输出{"Driver.age":35,"Car.type":"赛跑","Driver.name":"全爷","Car.name":"保时捷"}

本章小结

本章首先介绍了 Spring Boot 的 Web 开发支持,然后详细讲述了 Spring Boot 推荐使用的 Thymeleaf 模板引擎,包括 Thymeleaf 的基础语法、常用属性和国际化。同时,本章还介绍了 Spring Boot 对 JSON 数据的处理、文件上传下载、异常统一处理和对 JSP 的支持等 Web 应用开发的常用功能。

在线测试

习题

一、单选题

1. 以下关于 Thymeleaf 模板引擎常用标准表达式的说法,错误的是(　　)。
   A. 变量表达式♯{…}主要用于获取上下文中的变量值
   B. 使用 th:text="${♯locale.country}"动态获取当前用户所在国家信息
   C. 使用消息表达式♯{…}进行国际化设置时,还需要提供一些国际化配置文件
   D. 片段表达式~{…}用来标记一个片段模板,并根据需要移动或传递给其他模板

2. 以下关于 Spring Boot 整合 Thymeleaf 的相关配置说法,正确的是(　　)。
   A. spring.thymeleaf.cache 表示是否开启 Thymeleaf 模板缓存,默认为 false
   B. spring.thymeleaf.prefix 指定了 Thymeleaf 模板页面的存放路径,默认为 resources/
   C. spring.thymeleaf.suffix 指定了 Thymeleaf 模板页面的名称后缀,默认为.html
   D. spring.thymeleaf.encoding 表示模板页面变化格式,默认为 iso8859-1

3. 以下关于 Thymeleaf 模板引擎页面标签的说法,错误的是(　　)。
   A. th:each 用于元素遍历,类似 JSP 中的 c:forEach 标签
   B. th:value 用于属性值修改,指定标签属性值
   C. th:utext 用于指定标签显示的文本内容,对特殊标签进行转义
   D. th:href 用于设定链接地址

4. IE 不同版本 User-Agent 中出现的关键词不同,以下不属于 IE User-Agent 关键词的是(   )。
   A. MSIE   B. Mozilla   C. Edge   D. Trident
5. Thymeleaf 支持处理多种模板视图,不包括(   )。
   A. CSS   B. XML   C. JS   D. EXE
6. 在 Spring Boot 中使用路径扫描的方式整合内嵌式 Servlet 三大组件时,不包括的注解或属性是(   )。
   A. @WebServlet 注解   B. @EnableWebMvc 注解
   C. @ServletComponentScan 注解   D. value 属性

二、多选题

1. 以下关于 Spring Boot 整合 Spring MVC 框架实现 Web 开发中文件上传功能的相关说法,错误的是(   )。
   A. 实现文件上传功能,还需要提供文件上传相关依赖
   B. 必须在配置文件中对文件上传功能进行配置
   C. 多文件上传处理类中的方法中必须由 MultipartFile[ ]类型参数进行多文件接收
   D. spring.servlet.multipart.max-file-size 用来设置所有上传文件的大小限制,默认值为 10MB
2. 以下关于 Thymeleaf 主要标准表达式语法及说明,正确的是(   )。
   A. Thymeleaf 模板页面中的 th:text="${#locale.country}"动态获取当前用户所在国家信息
   B. ${#object.firstName}使用 Thymeleaf 模板提供的内置对象 object 获取当前上下文对象中的 firstName 属性值
   C. <div th:insert="~{thymeleafDemo::title}"></div>中的 title 为引入的模板名称
   D. 使用 th:insert 或 th:replace 属性可以插入 Thymeleaf 模板片段
3. Spring Boot 框架整合 MVC 实现 Web 开发支持的前端模板引擎包括(   )。
   A. Mustache   B. FreeMarker   C. Thymeleaf   D. Groovy

三、判断题(对的打"√",错的打"×")

1. 在 Spring Boot 项目的 classpath:/static/目录下编写一个 index.html 页面,可以作为 Spring Boot 默认欢迎页。(   )
2. 在 Spring Boot 中进行中文名文件下载处理时,如果内核信息是 IE,则转码为 ISO-8859-1 进行处理。(   )
3. Thymeleaf 支持处理 6 种模板视图,包括 HTML、XML、TEXT、JAVASCRIPT、CSS 和 RAW。(   )
4. 使用 data-th-* 属性定制 Thymeleaf 模板页面时,不需要引入 Thymeleaf 标签。(   )
5. Thymeleaf 是适用于 Web 和独立环境的现代服务器端 Java 模板引擎。(   )

四、填空题

1. Spring Boot 为整合 Spring MVC 框架实现 Web 开发,支持静态项目首页_____。

2. Spring Boot 整合 Spring MVC 实现 Web 开发,需要引入依赖启动器_____。

3. 在 Spring Boot 中,使用组件注册方式整合内嵌 Servlet 容器的 Filter 组件时,只需将自定义组件通过_____类注册到容器中即可。

4. Spring Boot 为整合 Spring MVC 框架实现 Web 开发,内置了 ContentNegotiating-ViewResolver 和_____两个视图解析器。

5. 消息表达式_____主要用于 Thymeleaf 模板页面国际化内容的动态替换和展示。

第4章

Spring Boot数据访问

本章学习目标

- 掌握 Spring Boot 整合 JDBC。
- 掌握 Spring Boot 整合 MyBatis。
- 掌握 Spring Boot 整合 JPA。
- 掌握数据缓存 Cache。

自党的十八大以来,网络与信息安全保障体系达到了前所未有的高度。根据党的二十大报告,要加强个人信息保护,持续强化数据等安全保障体系建设。

Spring Data 是 Spring 访问数据库的一系列解决方案,其中包含大量关系型数据库和非关系型数据库的数据访问解决方案。Spring Boot 在简化项目开发和实现自动化配置的基础上,对关系型数据库的访问操作提供了非常好的整合支持。

本章将针对 Spring Boot 的数据访问进行讲解,并进而体会数据安全的重要性。

4.1 Spring Boot 整合 JDBC

在开发中,通常会涉及对数据库的数据进行操作,Spring Boot 在简化项目开发和实现自动化配置的基础上,对关系型数据库和非关系型数据库的访问操作都提供了非常好的整合支持。

用 Spring Boot 框架开发时,常用的 ORM 框架有 JDBC、MyBatis、Hibernate、JPA 等,接下来学习 Spring Boot 整合 JDBC 的实现。

4.1.1 Spring Data 简介

无论是 SQL 还是 NoSQL,Spring Boot 默认采用整合 Spring Data 的方式进行统一处理,添加大量自动配置,屏蔽了很多设置,引入各种 xxxTemplate、xxxRepository 来简化对数据访问层的操作。因此,只需要进行简单的设置即可,需要用什么数据访问,就引入相关的启动器进行开发。Spring Data 的常用启动器如表 4-1 所示。

表 4-1　Spring Data 的常用启动器

依 赖 包	作　　用
spring-boot-starter-jdbc	支持 Tomcat JDBC 连接池
spring-boot-starter-batch	支持 Spring Batch，包括 HSQLDB 数据库
spring-boot-starter-data-jpa	支持 JPA，包括 spring-data-jpa、spring-orm、Hibernate
spring-boot-starter-data-mongodb	支持 MongoDB 数据，包括 spring-data-mongodb
spring-boot-starter-data-elasticsearch	支持 ElasticSearch 搜索和分析引擎，包括 spring-data-elasticsearch
spring-boot-starter-data-gemfire	支持 GemFire 分布式数据存储，包括 spring-data-gemfire
spring-boot-starter-data-rest	支持通过 REST 连接 Spring Data 数据仓库，包括 spring-data-rest-webmvc
spring-boot-starter-data-solr	支持 Apache Solr 搜索平台，包括 spring-data-solr
spring-boot-starter-redis	支持 Redis 键值存储数据库，包括 spring-redis

视频讲解

4.1.2　整合 JDBC Template

Spring 框架针对数据库开发中的应用提供了 JdbcTemplate 类，该类是 Spring 对 JDBC 支持的核心，它提供了所有对数据库操作功能的支持。

Spring 框架提供的 JDBC 支持主要由 4 个包组成，分别是 core(核心包)、object(对象包)、dataSource(数据源包)和 support(支持包)，org.springframework.jdbc.core.JdbcTemplate 类就包含在核心包中。作为 Spring JDBC 的核心，JdbcTemplate 类中包含了所有数据库操作的基本方法。

JdbcTemplate 主要提供以下几类方法。

- execute 方法：可以用于执行任何 SQL 语句，一般用于执行 DDL 语句。
- update 方法及 batchUpdate 方法：update 方法用于执行新增、修改、删除等语句，batchUpdate 方法用于执行批处理相关语句。
- query 方法及 queryForXXX 方法：用于执行查询相关语句。
- call 方法：用于执行存储过程、函数相关语句。

在 Spring Boot 应用中，如果使用 JdbcTemplate 操作数据库，那么只需要在 pom.xml 文件中添加 spring-boot-starter-jdbc 模块，即可通过 @Autowired 注解依赖注入 JdbcTemplate 对象，然后调用 JdbcTemplate 提供的方法操作数据库。

下面通过示例讲解如何在 Spring Boot 应用中使用 JdbcTemplate 操作数据库。

(1) 以 3.4 节中的 springboot0304 项目为基础，创建项目 springboot0401。

(2) 选择相应依赖。

使用 Spring Initializr 方式创建项目，在 Dependencies 依赖选择中除了选择 Web 模块下的 Web 场景依赖和 Template Engines 模块下的 Thymeleaf 场景依赖外，还需要选择 SQL 模块下的 JDBC API 场景依赖和 MySQL Driver 场景依赖，然后根据提示完成项目的创建。引入的场景依赖效果如图 4-1 所示。

项目创建成功后，在 pom.xml 文件中就添加了 MySQL 连接器和 spring-boot-starter-jdbc 模块，代码如下。

```
<!-- 引用 JdbcTemplate 依赖 -->
<dependency>
    <groupId>org.springframework.boot</groupId>
```

```xml
        <artifactId>spring-boot-starter-jdbc</artifactId>
</dependency>
<!-- 这里使用 MySQL 5.X 版本 -->
<dependency>
    <groupId>mysql</groupId>
    <artifactId>mysql-connector-java</artifactId>
    <version>5.1.30</version>
</dependency>
```

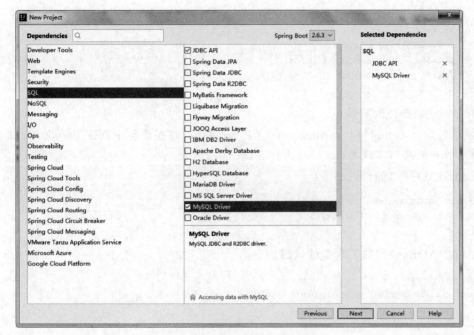

图 4-1　引入的场景依赖效果图

（3）配置数据源。

在 application.properties 文件中进行数据库连接配置，代码如下。

```
spring.datasource.url=jdbc:mysql://localhost:3306/book
spring.datasource.username=root
spring.datasource.password=123456
#MySQL 的版本为 8.X
#spring.datasource.driver-class-name=com.mysql.cj.jdbc.Driver
#MySQL 的版本为 5.X
spring.datasource.driver-class-name=com.mysql.jdbc.Driver
```

（4）创建数据表。

```sql
CREATE TABLE `bookinfo` (
  `ISBN` varchar(20) NOT NULL DEFAULT '' COMMENT '图书的 ISBN 发行号',
  `TITLE` varchar(50) NOT NULL DEFAULT '' COMMENT '书名',
  `AUTHOR` varchar(30) NOT NULL DEFAULT '' COMMENT '作者',
  `PUBLISHER` varchar(30) NOT NULL DEFAULT '' COMMENT '出版社',
  `PUBLISH_DATE` varchar(20) DEFAULT NULL COMMENT '出版日期',
  `PRICE` decimal(5,1) DEFAULT NULL COMMENT '单价',
  `INTRODUCE` varchar(255) NOT NULL DEFAULT '' COMMENT '简介',
  PRIMARY KEY (`ISBN`)
) ENGINE=MyISAM DEFAULT CHARSET=gbk;
```

（5）创建实体类。

在 java 目录下添加 it.com.boot.springboot04_01.po 包，并创建 Book 实体类，用作数据对象的封装。

```java
public class Book {
    private String title;
    private String author;
    private String publisher;
    private String publish_date;
    private double price;
    private String introduce;
    private String 综合业务数字网;
    //省略 get 和 set 方法
}
```

（6）创建数据访问层。

创建名为 it.com.boot.springboot04_01.dao 的包，并在该包中创建 BookDAO 接口和 BookDAOImpl 接口实现类。

BookDAO 接口的代码如下：

```java
public interface BookDAO {
    List<Book> findAllBooks();
}
```

BookDAOImpl 接口实现类的代码如下：

```java
@Repository
public class BookDAOImpl implements BookDAO {
    private String sql;
    @Autowired
    private JdbcTemplate jdbcTemplate;
    @Override
    public List<Book> findAllBooks() {
        sql = "select * from bookinfo ";
        RowMapper<Book> rowmapper = new BeanPropertyRowMapper<>(Book.class);
        List<Book> books = jdbcTemplate.query(sql, rowmapper);
        return books;
    }
}
```

（7）创建业务层。

创建名为 it.com.boot.springboot04_01.service 的包，并在该包中创建 BookService 接口和 BookServiceImpl 接口实现类。

BookService 接口的代码如下：

```java
public interface BookService {
    List<Book> findAllBooks();
}
```

BookServiceImpl 接口实现类的代码如下：

```java
@Service
public class BookServiceImpl implements BookService {
    @Autowired
```

```
    private BookDAO dao;
    @Override
    public List < Book > findAllBooks() {
        return dao.findAllBooks();
    }
}
```

（8）创建控制器类 BookController。

```
@Controller
public class BookController {
    @Autowired
    private BookService service;
    @RequestMapping("/login")
    public String findAll(Model model)
    {
        Calendar calendar = Calendar.getInstance();
        SimpleDateFormat dateFormat = new SimpleDateFormat("yyyy-MM-dd");
        List bookList = service.findAllBooks();
        model.addAttribute("dateString", dateFormat.format(calendar.getTime()));
        model.addAttribute("bookList", bookList);
        return "booklist";
    }
}
```

（9）运行程序，效果如图 3-17 所示。

4.1.3　数据库连接池 Druid

使用数据库连接池主要是考虑到程序与数据库建立连接的性能。创建一个新的数据库是一个很耗时的过程，在使用完之后，可能还需要不断地释放建立的连接，对资源的损耗很大。

使用数据库连接池会创建固定数量的数据库连接，需要用的时候使用即可。当然，这样做的一个缺点是，可能某些时候完全没有数据库请求，但是也保持了数据库的最小连接数，同样浪费了资源。不过，相对于完全不采用数据库连接池的资源浪费情况，这种方式还是很有优势的。

Druid 是阿里巴巴开源平台上一个数据库连接池的实现，并已经在该平台部署了 600多个应用。它结合了 C3P0、DBCP 等数据连接池的优点，同时加入了日志监控，可以很好地监控 DB 池连接和 SQL 的执行情况，可以说是针对监控而生的数据连接池。

Spring Boot 2.0 以上版本默认使用 Hikari 数据源，可以说 Hikari 与 Driud 都是当前Java Web 上最优秀的数据源。下面重点介绍 Spring Boot 如何集成 Druid 数据源。

（1）以项目 springboot0401 为基础，创建 springboot0402。

（2）打开 pom.xml 文件，添加对 Druid 的依赖引用。

```xml
< dependency >
    < groupId > com.alibaba </ groupId >
    < artifactId > druid-spring-boot-starter </ artifactId >
    < version > 1.1.10 </ version >
</ dependency >
```

（3）修改 Spring Boot 的默认数据源。

Spring Boot 的默认数据源是 org.apache.tomcat.jdbc.pool.DataSource。因为这里使用的是 Druid，所以需要修改 spring.datasource.type 为 druid。

```
#在添加 Druid 依赖时，由于直接选择适配 Spring Boot 开发的 Druid 启动器 druid-spring-boot-
starter,因此可以不需要再进行此项的配置，项目就会自动识别该数据源
spring.datasource.type = com.alibaba.druid.pool.DruidDataSource
#连接池的设置
#初始化时建立物理连接的个数
spring.datasource.initialSize: 5
#最小连接池数量
spring.datasource.minIdle: 5
#最大连接池数量 maxIdle 已经不再使用
spring.datasource.maxActive: 20
#获取连接时的最大等待时间，单位为 ms
spring.datasource.maxWait: 60000
#既作为检测的间隔时间，又作为 testWhileIdle 执行的依据
spring.datasource.timeBetweenEvictionRunsMillis: 60000
#在销毁线程时检测，当前连接的最后活动时间和当前时间差大于该值时，关闭当前连接
spring.datasource.minEvictableIdleTimeMillis: 300000
#在申请连接时检测，如果空闲时间大于 timeBetweenEvictionRunsMillis，执行 validationQuery 检
测连接是否有效
spring.datasource.testWhileIdle: true
#申请连接时会执行 validationQuery 检测连接是否有效，开启会降低性能，默认为 true
spring.datasource.testOnBorrow: false
#归还连接时会执行 validationQuery 检测连接是否有效，开启会降低性能，默认为 true
spring.datasource.testOnReturn: false
#是否缓存 preparedStatement，建议开启 mysql5.5 +
spring.datasource.poolPreparedStatements: true
```

在配置项中指定了各个参数后，在连接池内部对这些参数的使用过程如下。数据库连接池在初始化的时候会创建 initialSize 个连接，当有数据库操作时，会从池中取出一个连接。如果当前池中正在使用的连接数等于 maxActive，则会等待一段时间，等待其他操作释放掉某一个连接，如果这个等待时间超过了 maxWait，则会报错；如果当前正在使用的连接数没有达到 maxActive，则判断当前是否有空闲连接，如果有则直接使用空闲连接，如果没有则新建立一个连接。在连接使用完毕后，不是将其物理连接关闭，而是将其放入池中并等待其他操作复用。

连接池内部有机制判断，如果当前的总连接数小于 miniIdle，则会建立新的空闲连接，以保证连接数达到 miniIdle。如果当前连接池中某个连接在空闲了 timeBetweenEvictionRunsMillis 时间后仍然没有使用，则被物理性地关闭掉。有些数据库在连接时有超时限制（MySQL 连接在 8 小时后断开），或者由于网络中断等原因，连接池的连接会出现失效的情况，这时可以设置一个参数为 true 的 testWhileIdle。需要注意的是，这里的"while"应该翻译成"如果"，即 testWhileIdle 写为 testIfIdle 更易于理解，其含义为在获取连接时，如果检测到当前连接不活跃的时间超过了 timeBetweenEvictionRunsMillis，则手动检测一下当前连接的有效性，在保证确实有效后才加以使用。在检测活跃性时，如果当前的活跃时间大于 minEvictableIdleTimeMillis，则认为需要关闭当前连接。当然，为了保证绝对的可用性，也可以使用参数为 true 的 testOnBorrow（即在每次获取 Connection 对象时都检测其可用性），

不过这样会影响性能。

(4) 手动配置属性。

在配置好数据源后，数据源从默认的 tomcat.pool 切换到 Druid，可是 Druid 的属性（如 initialSize、minIdle）还未生效。因为在 DataSourceProperties.class 下没有对应的匹配项，所以还需要手动配置。

创建 it.com.boot.springboot04_02.config 包，并在该包下创建一个自定义配置类，对 Druid 数据源属性值进行注入。

```
@Configuration
public class DruidConfig {
    @Bean
    @ConfigurationProperties(prefix = "spring.datasource")
    public DataSource getDataSource() {
        return new DruidDataSource();
    }
}
```

(5) 注入 DataSource。

在数据源切换后，在测试类中注入 DataSource 并进行获取。测试运行结果如图 4-2 所示，从图中可见配置参数已经生效。

```
@SpringBootTest
class Springboot0402ApplicationTests {
    @Autowired
    private DataSource dataSource;
    @Test
    void contextLoads() throws SQLException {
        System.out.println(dataSource.getClass());
        Connection connection = dataSource.getConnection();
        System.out.println(connection);
        DruidDataSource druidDataSource = (DruidDataSource) dataSource;
        System.out.println("数据源最大连接数：" + druidDataSource.getMaxActive());
        System.out.println("数据源初始化连接数：" + druidDataSource.getInitialSize());
        connection.close();
    }
}
```

```
2022-03-28 16:34:24.281  INFO 15544 --- [           main] i.c.b.s.Springboot0402ApplicationTests
 class com.alibaba.druid.pool.DruidDataSource
2022-03-28 16:34:25.786  INFO 15544 --- [           main] com.alibaba.druid.pool.DruidDataSource
com.mysql.jdbc.JDBC4Connection@2fdf17dc
数据源最大连接数：20
数据源初始化连接数：5
2022-03-28 16:34:25.887  INFO 15544 --- [ionShutdownHook] com.alibaba.druid.pool.DruidDataSource
```

图 4-2 测试结果

4.2 Spring Boot 整合 MyBatis

视频讲解

在使用 Spring Boot 的过程中，常用的持久化解决方案主要有两种，一种是 MyBatis 框架，另一种是 Spring Data JPA。

Spring Data JPA 和 MyBatis 最大的区别在于 Spring Data JPA 是 Spring 的直系衍生框架,这一点从名字的命名方式上也能看出来。在使用 MyBatis 时可以发现,MyBatis 依赖的 artifactId 是 mybatis-spring-boot-starter,而 Spring Data JPA 依赖的 artifactId 却是 spring-boot-starter-data-jpa,从这个关键字(mybatis、jpa)的顺序上就能看出来谁才是直系衍生的。

虽然不是直系衍生的,但是 MyBatis 凭借自己轻巧灵活的特性(易上手、动态 SQL 等),赢得了广大开发者的喜爱。MyBatis 是一款优秀的持久层框架,它支持定制化 SQL、存储过程和高级映射。MyBatis 避免了几乎所有的 JDBC 代码和手动设置参数及获取结果集。MyBatis 可以使用简单的 XML 或注解来配置和映射原生信息,将接口和 Java 的 POJOs(Plain Old Java Objects,普通的 Java 对象)映射成数据库中的记录。

下面通过示例讲解 Spring Boot 整合 MyBatis 的过程。

(1) 创建项目。

以项目 springboot0401 为基础,创建项目 springboot0403。首先创建一个基本的 Spring Boot 工程,添加 Web 依赖、MyBatis 依赖和 MySQL 驱动依赖,如图 4-3 所示。

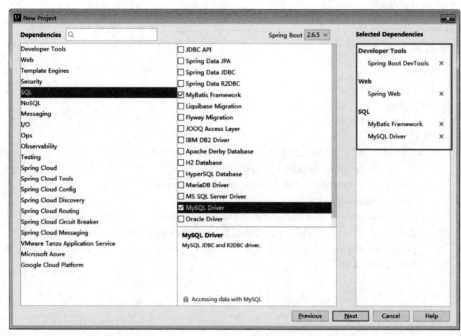

图 4-3 创建 MyBatis 整合项目

(2) 添加依赖。

在项目创建成功后,添加 Druid 依赖,并且锁定 MySQL 驱动版本,完整的依赖如下。

```
<!-- 这里使用MySQL 5.X版本 -->
<dependency>
    <groupId>mysql</groupId>
    <artifactId>mysql-connector-java</artifactId>
    <version>5.1.30</version>
</dependency>
<!-- 添加Druid -->
<dependency>
```

```xml
        <groupId>com.alibaba</groupId>
        <artifactId>druid-spring-boot-starter</artifactId>
        <version>1.1.10</version>
</dependency>
<!-- 添加Mybatis -->
<dependency>
        <groupId>org.mybatis.spring.boot</groupId>
        <artifactId>mybatis-spring-boot-starter</artifactId>
        <version>1.3.2</version>
</dependency>
```

通常在集成一些 Spring Boot 提供支持的技术时，所添加的依赖都是以 spring-boot-starter 开头的 spring-boot-starter-xxx，但是刚才添加的 MyBatis 依赖却是以 mybatis 开头的 mybatis-spring-boot-starter。这其实是因为 Spring Boot 默认不支持 MyBatis，它默认支持的是它自己的持久层框架 JPA，由于 Spring Boot 是大势所趋，因此 MyBatis 就主动去迎合 Spring Boot 生态，自己开发了 MyBatis 的 stater。凡是格式为 xxx-spring-boot-starter 的依赖，都是 Spring Boot 没有主动提供支持的技术。

mybatis-spring-boot-starter 依赖将会提供如下功能：

- 自动检测现有的 DataSource。
- 创建并注册 SqlSessionFactory 的示例，该示例使用 SqlSessionFactoryBean 将该 DataSource 作为输入进行传递。
- 创建并注册从 SqlSessionFactory 中获取的 SqlSessionTemplate 的示例。
- 自动扫描 Mappers，将示例连接到 SqlSessionTemplate 并将其注册到 Spring 上下文中，以便将它们注入 Bean 中。

也就是说，在使用了该 starter 后，只需要定义一个 DataSource 即可（在 application. Properties 中配置），它会自动创建使用该 DataSource 的 SqlSessionFactoryBean 和 SqlSessionTemplate。然后自动扫描 Mappers，将示例连接到 SqlSessionTemplate，并注册到 Spring 上下文中。

（3）在 application.properties 中添加数据库连接配置。

```
spring.datasource.url=jdbc:mysql://localhost:3306/book
spring.datasource.username=root
spring.datasource.password=123456
spring.datasource.driver-class-name=com.mysql.jdbc.Driver
```

在 pom.xml 中添加第三方数据源 Druid 依赖时，由于直接选了适配 Spring Boot 开发的 Druid 启动器 druid-spring-boot-starter，因此可以不需要再进行其他配置，项目就会自动识别该数据源。有时可能会使用独立的 Druid 依赖文件，这时就必须在全局配置文件中额外添加 spring.datasource.type=com.alibaba.druid.pool.DruidDataSource 配置，这样项目才会识别配置的 Druid 数据源。

4.2.1 使用配置文件的方式整合 MyBatis

对数据库进行操作就需要编写 SQL 语句，这里将沿用 Spring 整合 MyBatis 的方式，将 SQL 语句放在 XML 配置文件中，具体操作步骤如下。

(1) 配置 MyBatis。

```
#mapper 配置文件
  mybatis.mapper-locations=classpath:mapper/*.xml
#实体类的别名
  mybatis.type-aliases-package=it.com.boot.springboot04_03.po
#开启驼峰命名
  mybatis.configuration.map-underscore-to-camel-case=true
```

① mybatis.mapper-locations 的作用是扫描 Mapper 接口对应的 XML 文件,上述代码中扫描的是 resources 目录下的 mapper 文件夹中所有以 xml 结尾的文件。

② mybatis.type-aliases-package 属性指定实体类扫描路径,让 MyBatis 自动扫描到自定义的实体类。如果不指定路径,在 MyBatis 的 mapper.xml 文件中,resultType 的 type 或 parameterType 会返回自定义实体类,此时需要用完整的类路径和类名来指定这些实体。

③ mybatis.configuration.map-underscore-to-camel-case 的作用是将带有下画线的表字段映射为驼峰格式的实体类属性。MySQL 定义字段一般用下画线表示,比如 student_name,而在实体类中通常把对应的属性名定义为 studentName,这样 SQL 语句查询返回的字段和映射实体类属性不一致,会导致在写好 SQL 查询语句后,发现查询出的对象字段都为 null。因此,在添加了该项配置后,在开发中只需要根据查询返回的字段创建好实体类就可以了。

(2) 创建数据访问接口。

创建 it.com.boot.springboot04_03.dao 包,并在该包下创建一个操作 bookinfo 表的 BookInfoDao 接口。

```
@Mapper
public interface BookInfoDao {
    List<Book> findAllBooks();
    Book findBookByIsbn(String isbn);
    Integer addBook(Book book);
    Integer updateBook(Book book);
    Integer delBookByIsbn(String isbn);
}
```

这是 MyBatis 的 Mapper 映射接口,不需要使用@Repository 注解数据访问层,Spring Boot 默认将@Mapper 视为数据访问层。

(3) 创建 Mapper 映射文件。

在 src/main/resources 目录下创建名为 mapper 的包,并在该包下创建 SQL 映射文件 BookMapper.xml,将接口中方法对应的 SQL 直接写在 XML 文件中,实现数据库的相关操作。

```xml
<mapper namespace="it.com.boot.springboot04_03.dao.BookInfoDao">
  <select id="findAllBooks" resultType="book">
      select * from bookinfo
  </select>
  <select id="findBookByIsbn" parameterType="string" resultType="book">
      select * from bookinfo where isbn=#{isbn}
  </select>
  <insert id="addBook" parameterType="book">
      insert into bookinfo values
```

```xml
        (#{ISBN},#{title},#{author},#{publisher},#{publishDate},#{price},#{introduce})
    </insert>
    <update id = "updateBook" parameterType = "book">
        update bookinfo
        <set>
            <if test = "author!= null and author!= ''"> author = #{author},</if>
            <if test = "introduce!= null and introduce!= ''"> introduce = #{introduce},</if>
        </set>
        <where> ISBN = #{ISBN}</where>
    </update>
    <delete id = "delBookByIsbn" parameterType = "string">
        delete from bookinfo where ISBN = #{isbn}
    </delete>
</mapper>
```

BookMapper.xml 有以下两种存放方式。

① 第一种方式是直接放在 BookInfoDao 接口所在的包下面。放在这里的 BookMapper.xml 会被自动扫描到，但是有另外一个 Maven 带来的问题，就是 Java 目录下的 xml 资源在项目打包时会被忽略掉。所以，如果将 BookMapper.xml 放在这里，需要在 pom.xml 文件中再添加相应的配置，避免打包时 Java 目录下的 XML 文件被自动忽略掉。

② 第二种方式是将 BookMapper.xml 直接放在 resources 目录下，这样就不用担心打包时被忽略的问题。此时 mapper 文件不能被自动扫描到，需要添加额外配置，之前在全局配置文件中添加的 mybatis.mapper-locations 属性可以起到作用，这样 mapper 就可以正常使用了。

（4）创建业务层和控制器。

具体代码和 springboot04_01 中的一致，这里不再赘述。

（5）运行程序，效果如图 3-17 所示。

4.2.2 使用注解的方式整合 MyBatis

通过 4.2.1 节的实例可以看到，相对于 MyBatis 初期使用，即需要各种配置文件、实体类、Dao 层映射关联等，Spring Boot 以配置文件的方式整合 MyBatis 已经简化了很多。但是通过使用注解，可以进一步进行简化，不需要任何 XML 配置文件。下面介绍这种使用注解的方式，实现简单配置。

1. 基础注解

MyBatis 主要提供的 CRUD 注解如下。

- @Select
- @Insert
- @Update
- @Delete

增、删、改、查占据了绝大部分的业务操作，掌握这些基础注解的使用是很有必要的。例如，下面这段代码无须 XML 即可完成数据查询。

```java
@Mapper
public interface BookInfoDao {
@Select("select * from bookinfo")
```

```
    List<Book> findAllBooks();
}
```

使用过 Hibernate 的读者可能会好奇，这里为什么没有配置映射关系也能完成属性注入？在传统项目中使用过 MyBatis 的读者可能很快就能反应过来，这是因为在配置文件中开启了全局驼峰映射，Spring Boot 中同样能够做到，并且更为简单快捷。

虽然开启了全局驼峰映射，但如果字段不符合下画线转驼峰的规则，查询回来的实体对象属性将获取为 null。例如，上述 Book 对象属性 pdate 和对应的数据库字段 PUBLISH_DATE 不相同，则查询结果如下：

```
[
  {
  "title":"Java 程序设计",
  "author":"杜雷",
  "publisher":"清华大学出版社",
  "pdate":null,
  "price":23.9,
  "introduce":"关于 Java 程序设计方面的书",
  "isbn":"1202255889"
  }
  …
]
```

为了解决对象属性和字段驼峰不一致的问题，可以使用映射注解@Results 来指定映射关系。

2. 映射注解

MyBatis 主要提供的映射注解如下。

- @Results：用于填写结果集的多个字段的映射关系。
- @Result：用于填写结果集的单个字段的映射关系。
- @ResultMap：根据 ID 关联 XML 里面的<resultMap>。

例如，上面的 findAllBooks 方法，可以在查询 SQL 的基础上，指定返回结果集的映射关系，其中 property 表示实体对象的属性名，column 表示对应的数据库字段名。

```
@Mapper
public interface BookInfoDao {
    @Results(
        @Result(column = "PUBLISH_DATE", property = "pdate")
    )
    @Select("select * from bookinfo")
    List<Book> findAllBooks();
}
```

@Results 为结果映射配置，如果属性名称和表的字段名称一致的话，则不需要此配置也可以自动完成映射。

对数据库的查询结果如下：

```
[
  {
  "title":"Java 程序设计",
  "author":"杜雷",
  "publisher":"清华大学出版社",
```

```
    "pdate":"2018-1-1",
    "price":23.9,
    "introduce":"关于 Java 程序设计方面的书",
    "isbn":"1202255889"
    }
    …
]
```

3. 动态条件判断

如果想使用类似 XML 中的 if-else,需要用到<script></script>区间块标志。修改 BookInfoDao 文件,输入以下内容,以支持自定义 script 查询操作。

```
@Update("<script> update bookinfo" +
    " <set>" +
    " <if test = \"author!= null and author!= ''\"> author = #{author},</if>" +
    " <if test = \"introduce!= null and introduce!= ''\"> introduce = #{introduce},</if>" +
    " </set>" +
    " <where> ISBN = #{ISBN}</where></script>")
Integer updateBook(Book book);
```

4. 共享 Results 结果数据

在实际开发中,不同的方法可以返回相同的结果信息,这时就用到了共享 Results 结果数据。共享结果数据很简单,只需要修改@Results,在里面增加 id 属性(该 id 值是唯一的),后期在其他方法中使用即可。

修改 BookInfoDao 文件中的 findAllBooks 方法,在@Results 中增加 id 属性,代码如下。

```
@Results(id = "bookResult", value = {
    @Result(column = "PUBLISH_DATE", property = "pdate")}
)
@Select("select * from bookinfo")
List<Book> findAllBooks();
```

定义一个 id 为 bookResult 的返回全局结果集,修改 BookInfoDao 文件,输入以下内容,以支持共享结果查询操作。

```
@ResultMap("bookResult")
@Select(" select * from bookinfo where isbn = #{id}")
Book findBookByIsbn(@Param("id") String isbn);
```

注意此时使用的是@ResultMap 注解,该注解只有一个 value 属性,指定需要返回哪个数据结果集。

同时需要注意,在这个方法中使用了@Param 注解,它的作用是指定 SQL 语句中的参数,通常用于 SQL 语句中参数比较多的情况。在上面的示例中,使用@Param 注解给参数 isbn 命名为 id。此时 isbn 这个参数命名需要与@Select 注解中的 #{} 中名称一一对应,@Param 注解用于将参数 isbnde 值映射到 #{id} 中。

下面使用注解的方式改写 springboot0403,具体步骤如下。

(1) 以项目 springboot0403 为基础,创建 springboot0404。

(2) 创建实体类。

由于本实例特意要使用@Result 注解,因此在创建 book 的实体类中修改了出版日期的属性名称,这与 book 表中的出版日期所对应的字段不一致。

```java
public class Book {
    private String title;
    private String author;
    private String publisher;
    private String pdate; //出版日期
    private double price;
    private String introduce;
    private String ISBN;
//省略 get 和 set 方法
}
```

(3) 创建 Mapper 接口文件 BookInfoDao。

```java
@Mapper
public interface BookInfoDao {
    @Results(id = "bookResult", value = {
            @Result(column = "PUBLISH_DATE", property = "pdate")}
    )

    @Select("select * from bookinfo")
    List<Book> findAllBooks();

    @ResultMap("bookResult")
    @Select(" select * from bookinfo where isbn = #{id}")
    Book findBookByIsbn(@Param("id") String isbn);

    @Insert("INSERT into bookinfo VALUES(#{ISBN},#{title},#{author},#{publisher},#{pdate},#{price},#{introduce})")
    Integer addBook(Book book);

    @Update("<script> update bookinfo" +
            " <set>" +
            " <if test = \"author!= null and author!= ''\"> author = #{author},</if>" +
            " <if test = \"introduce!= null and introduce!= ''\"> introduce = #{introduce},</if>" +
            " </set>" +
            " <where> ISBN = #{ISBN}</where></script>")
    Integer updateBook(Book book);

    @Delete(" delete from bookinfo where ISBN = #{isbn}")
    Integer delBookByIsbn(String isbn);
}
```

(4) 运行程序,效果如图 3-17 所示。

4.3　Spring Boot 整合 JPA

JPA 是 Java Persistence API 的简称,中文名为 Java 持久层 API。

JPA 是 Sun 公司提出的 Java 持久化规范,其设计目标主要是为了简化现有的持久化开发工作和整合 ORM 技术。

JPA 使用 XML 文件或注解(JDK 5.0 或更高版本)来描述对象-关联表的映射关系,能够将运行期的实体对象持久化到数据库,它为 Java 开发人员提供了一种 ORM 工具来管理

Java 应用中的关系数据。

简单地说，JPA 就是为 POJO(Plain Ordinary Java Object)提供持久化的标准规范，即将 Java 的普通对象通过对象关系映射(Object Relational Mapping，ORM)持久化到数据库中。

由于 JPA 是在充分吸收了现有 Hibernate、TopLink、JDO 等 ORM 框架的基础上发展而来的，因而具有易于使用、伸缩性强等优点。

4.3.1 Spring Data JPA 简介

Spring Data JPA 是 Spring 基于 Spring Data 框架和 JPA 规范开发的一个框架，使用 Spring Data JPA 可以极大地简化 JPA 的写法，可以在几乎不用写实现的情况下实现对数据库的访问和操作。除了 CRUD 外，Spring Data JPA 还提供了对分页查询、自定义 SQL、查询指定多条记录、联表查询等功能的支持。

JPA 不是一种新的 ORM 框架，它的出现只是用于规范现有的 ORM 技术，它不能取代现有的 Hibernate、TopLink 等框架。相反，在采用 JPA 开发时，仍将使用到这些 ORM 框架，只是此时开发出来的应用不再依赖于某个持久化提供商，应用可以在不修改代码的情况下在任何 JPA 环境下运行，真正做到低耦合、可扩展的程序设计。Spring Data JPA、JSP 与 ORM 框架之间的关系如图 4-4 所示。

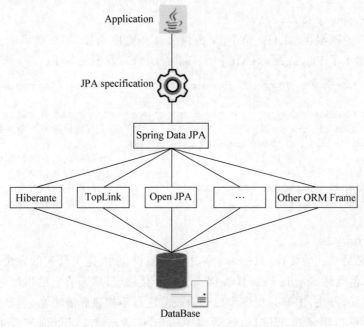

图 4-4　Spring Data JPA、JSP 与 ORM 框架之间的关系

Spring Data JPA 提供了如下几个常用的接口。

(1) CrudRepository 接口。

该接口提供了 11 个方法，基本上可以满足简单的 CRUD 操作及批量操作，其定义如下。

```
@NoRepositoryBean
public interface CrudRepository < T, ID extends Serializable > extends Repository < T, ID > {
```

```java
    < S extends T > S save(S entity);                        //保存
    < S extends T > Iterable < S > save(Iterable < S > entities);    //批量保存
    T findOne(ID id);                                        //根据 id 查询一个对象
    boolean exists(ID id);                                   //判断对象是否存在
    Iterable < T > findAll();                                //查询所有对象
    Iterable < T > findAll(Iterable < ID > ids);             //根据 id 列表查询所有对象
    long count();                                            //计算对象的总个数
    void delete(ID id);                                      //根据 id 删除对象
    void delete(T entity);                                   //删除单个对象
    void delete(Iterable <? extends T > entities);           //批量删除对象
    void deleteAll();                                        //删除所有对象
}
```

（2）PagingAndSortingRepository 接口。

PagingAndSortingRepository 接口继承了 CrudRepository 接口，只要继承了这个接口，Spring data JPA 就可以提供分页和排序的功能。该接口主要提供了两个方法，其中 T 是要操作的实体类，ID 是实体类主键的类型，其定义如下。

```java
@NoRepositoryBean
public interface PagingAndSortingRepository < T, ID extends Serializable > extends CrudRepository
< T, ID > {
    Iterable < T > findAll(Sort sort);                       //不带分页的排序
    Page < T > findAll(Pageable pageable);                   //带分页的排序
}
```

（3）JpaRepository 接口。

如果业务需要既提供 CRUD 操作，又提供分页和排序功能，那么就可以直接继承这个接口。该接口继承了 PagingAndSortingRepository 接口，其定义如下。

```java
public interface JpaRepository < T, ID extends Serializable > extends PagingAndSortingRepository
< T, ID > {
    List < T > findAll();                                    //查询所有对象,不排序
    List < T > findAll(Sort sort);                           //查询所有对象并排序
    < S extends T > List < S > save(Iterable < S > entities);    //批量保存
    void flush();                                            //强制缓存与数据库同步
    T saveAndFlush(T entity);                                //保存并强制同步
    void deleteInBatch(Iterable < T > entities);             //批量删除对象
    void deleteAllInBatch();                                 //删除所有对象
}
```

（4）Repository 接口。

这个接口是最基础的接口，只是一个标志性的接口，没有定义任何的方法。似乎该接口没有什么用处，但既然 Spring data JPA 提供了这个接口，自然是有它的用处。例如，有一部分不想对外提供的方法，比如只提供增加和修改方法而不提供删除方法，这是前面几个接口做不到的，此时可以继承这个接口，然后将 CrudRepository 接口里面相应的方法复制到 Repository 接口即可。

上述常用的 4 个接口，开发者到底该如何选择？其实依据很简单，根据具体的业务需求，进行选择其中，因为各接口之间并不存在功能强弱的问题。

视频讲解

4.3.2 简单条件查询

使用 Spring Data JPA 建立数据库访问层十分方便，只需要定义一个继承 JpaRepository 接

口的接口即可。因此,自定义的数据访问接口完全继承了 JpaRepository 的接口方法。但更重要的是,在自定义的数据访问接口中可以根据查询关键字定义查询方法,这些查询方法需要符合它的命名规则,一般是根据持久化实体类的属性名确定的。

在查询时,通常需要同时根据多个属性进行查询,且查询的条件也各式各样(大于某个值、在某个范围内等)。Spring Data JPA 为此提供了一些表达条件查询的关键字,如表 4-2 所示。

表 4-2　条件查询的关键字

关　键　字	示　　例	JPQL 代码段
And	findByLastnameAndFirstname	where x.lastname=?1 and x.firstname = ?2
Or	findByLastnameOrFirstname	where x.lastname=?1 or x.firstname = ?2
Is,Equals	findByFirstname, findByFirstnameIs, findByFirstnameEquals	where x.firstname=?1
Between	findByStartDateBetween	where x.startDate between ?1 and ?2
LessThan	findByAgeLessThan	where x.age < ?1
LessThanEqual	findByAgeLessThanEqual	where x.age ⇐ ?1
GreaterThan	findByAgeGreaterThan	where x.age > ?1
GreaterThanEqual	findByAgeGreaterThanEqual	where x.age >= ?1
After	findByStartDateAfter	where x.startDate > ?1
Before	findByStartDateBefore	where x.startDate < ?1
IsNull	findByAgeIsNull	where x.age is null
IsNotNull,NotNull	findByAge(Is)NotNull	where x.age not null
Like	findByFirstnameLike	where x.firstname like ?1
NotLike	findByFirstnameNotLike	where x.firstname not like ?1
StartingWith	findByFirstnameStartingWith	where x.firstname like ?1 (参数后加%,即以参数开头的模糊查询)
EndingWith	findByFirstnameEndingWith	where x.firstname like ?1 (参数前加%,即以参数结尾的模糊查询)
Containing	findByFirstnameContaining	where x.firstname like ?1 (参数两边加%,即包含参数的模糊查询)
OrderBy	findByAgeOrderByLastnameDesc	where x.age = ?1 order by x.lastname desc
Not	findByLastnameNot	where x.lastname <> ?1
In	findByAgeIn(Collection<Age> ages)	where x.age in ?1
NotIn	findByAgeNotIn(Collection<Age> age)	where x.age not in ?1
True	findByActiveTrue()	where x.active = true
False	findByActiveFalse()	where x.active = false
IgnoreCase	findByFirstnameIgnoreCase	where UPPER(x.firstame) = UPPER(?1)

下面通过示例讲解在 Spring Boot Web 应用中如何使用 Spring Data JPA 进行简单条件查询。

1. 以项目 springboot0401 为基础,创建项目 springboot0405

2. 添加依赖

添加 Thymeleaf、Spring Data JPA 依赖。在创建实体类时,为了省去代码中大量的

get、set 和 toString 等方法,这里可以勾选 Lombok 依赖,如图 4-5 所示。

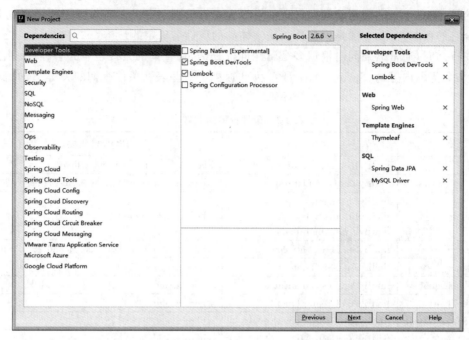

图 4-5　添加 Spring Data JPA 依赖

项目创建成功后,在 pom.xml 文件中就添加了 Spring Data JPA 模块依赖,代码如下。

```xml
<dependency>
    <groupId>org.springframework.boot</groupId>
    <artifactId>spring-boot-starter-data-jpa</artifactId>
</dependency>
<dependency>
    <groupId>org.projectlombok</groupId>
    <artifactId>lombok</artifactId>
    <optional>true</optional>
</dependency>
```

3. 配置 JPA 持久化信息

```
# JPA 持久化配置
# 是否开启 JPA Repositories,默认为 true
spring.data.jpa.repositories.enabled=true

# JPA 数据库类型,默认可以自动检测
spring.jpa.database=mysql

# 是否使用 JPA 初始化数据库,可以在启动时生成 DDL 创建数据库表,默认为 false
spring.jpa.generate-ddl=false

# 指定自动创建、更新数据库表等配置,update 表示如果数据库中存在持久化类对应的表则不创建,
不存在则创建
spring.jpa.hibernate.ddl-auto=update

# Hibernate 操作时显示真实的 SQL,默认为 false
spring.jpa.show-sql=true
```

4. 创建持久化实体类 Book

```
@Entity
@Table(name = "bookinfo")
@Data
public class Book implements Serializable{
  private String title;
  private String author;
  private String publisher;
  @Column(name = "PUBLISH_DATE")
  private String pdate;
  private Integer price;
  private String introduce;
  @Id
  private String ISBN;
}
```

上述代码定义了一个 Spring Data JPA 实体类 Book，并将该类与数据表 bookinfo 进行映射，下面针对其中用到的注解进行简要说明。

(1) @Entity。

@Entity 注解用于实体类声明语句之前，指出该 Java 类为实体类，将映射到指定的数据库表。在默认情况下，数据表的名称就是首字母小写的类名，如声明一个实体类 Customer，它将映射到数据库中的 customer 表上。当然，也可以使用 name 属性指定映射的表名。

(2) @Table。

① 当实体类与其映射的数据库表名不同名时需要使用 @Table 注解说明，该注解与 @Entity 注解并列使用，置于实体类声明语句之前，可写为单独语句行，也可与声明语句同行。

② @Table 注解的常用选项是 name，用于指明数据库的表名。

(3) @Id。

① @Id 注解用于声明一个实体类的属性映射为数据库的主键列。该属性通常置于属性声明语句之前，可与声明语句同行，也可写在单独行上。

② @Id 注解也可置于属性的 get 方法之前。

(4) @GeneratedValue。

① @GeneratedValue 用于注解主键的生成策略，通过 strategy 属性指定。在默认情况下，JPA 自动选择一个最适合底层数据库的主键生成策略，如 SqlServer 对应 identity、MySQL 对应 auto increment 等。

② 在 javax.persistence.GenerationType 中定义了以下几种可供选择的策略。

- IDENTITY：采用数据库 ID 自增长的方式来自增长键字段，Oracle 不支持这种方式。
- AUTO：默认选项，JPA 自动选择合适的策略。
- SEQUENCE：通过序列产生主键，通过 @SequenceGenerator 注解指定序列名，MySQL 不支持这种方式。
- TABLE：通过表产生主键，框架借由表模拟序列产生主键，使用该策略可以使应用更易于数据库移植。

(5) @Column。

① 当实体的属性与其映射的数据库表的列不同名时需要使用@Column注解说明,该注解通常置于实体的属性声明语句之前,可与@Id注解一起使用。

② @Column注解的常用属性是name,用于设置映射数据库表的列名。此外,该注解还包含其他多个属性,如unique、nullable、length等。

③ @Column注解也可置于属性的get方法之前。

(6) @Transient。

@Transient注解表示这个属性并非数据库表的映射字段,即表示非持久化属性。

5. 创建数据访问层

在dao包下新建接口类,将其命名为BookRepository。该接口要扩展JpaRepository<Book,String>接口,其中泛型传入要操作的实体类和主键类型。

```
public interface BookRepository extends JpaRepository<Book,String> {
    Book findByIsbn(String isbn);
}
```

在这里直接继承JpaRepository,其中已经有很多现成的方法(如内置的增、删、改、查方法等),不用自己写SQL语句,直接调用即可,这也是JPA的一大优点。

6. 创建业务层

```
public interface BookService {
    //查询所有图书
    List<Book> findAllBooks();
    //通过isbn查询图书
    Book findBookByIsbn(String isbn);
    //增加图书
    Book addBook(Book book);
    //修改图书
    Book updateShop(Book book);
    //通过isbn删除图书
    void delShop(String isbn);
}
@Service
public class BookServiceImpl implements BookService {
@Autowired
    private BookRepository dao;
    @Override
    public List<Book> findAllBooks() {
        return dao.findAll();
    }
    @Override
    public Book findBookByIsbn(String isbn) {
        return dao.findByISBN(isbn);
    }
    @Override
    public Book addBook(Book book) {
        return dao.save(book);
    }
    @Override
    public Book updateShop(Book book) {
```

```
            return dao.save(book);
        }
        @Override
        public void delShop(String isbn) {
            dao.deleteById(isbn);
        }
}
```

至此,JPA 最基本的使用就完成了。接口不需要实现,此时已经可以使用了。那么 JPA 是通过什么规则来根据方法名生成 SQL 查询语句呢?

其实 JPA 在这里遵循 Convention over configuration(约定优于配置)的原则,依据 Spring 和 JPQL 定义的方法命名。Spring 提供了一套可以通过命名规则进行查询构建的机制,这套机制会过滤一些关键字,如 find…By、read…By、query…By、count…By 和 get…By 等。系统会根据关键字将命名解析成两个子语句,第一个 By 是区分这两个子语句的关键词。By 之前的子语句是查询子语句(指明返回要查询的对象),后面的部分是条件子语句。如果关键字直接为 find By…,则返回的就是定义 Respository 时指定的领域对象集合。此时,JPQL 还定义了丰富的关键字,如 and、or、between 等。

7. 创建并运行控制层

BookController 控制器与项目 springboot0401 中的控制器代码相同,此处不再赘述。运行程序,效果如图 3-17 所示。

4.3.3 关联查询

在 Spring 工程中创建实体对象时,可以通过 JPA 的 @Entity 标识实体与数据库表的对应关系,通过 @Column 标识数据库字段。此外,还有标识两个实体间关系的注解:@OneToOne、@OneToMany、@ManyToOne 和 @ManyToMany,分别标识一对一、一对多、多对一和多对多关系。下面主要讲解 @OneToOne、@OneToMany、@ManyToOne 和 @ManyToMany 的使用,并对修饰关系维护字段的 @JoinColumn 进行简单介绍。

1. @OneToOne

@OneToOne,表示一对一的映射关系。例如,一个账号对应一个用户,一个实体用来描述账号的信息(账号、密码、账号是否可用、账号对应的角色等),另一个实体用来描述用户的信息(昵称、性别、国籍等)。

视频讲解

```
public @interface OneToOne {
    java.lang.Class targetEntity() default void.class;
    javax.persistence.CascadeType[] cascade() default {};
    javax.persistence.FetchType fetch() default javax.persistence.FetchType.EAGER;
    boolean optional() default true;
    java.lang.String mappedBy() default "";
    boolean orphanRemoval() default false;
}
```

该注解有以下 6 个属性:
(1) targetEntity 表示关联目标实体类,指定类型后该属性可省略。
(2) cascade 表示关联关系中的级联操作权限,有以下 5 种权限。
• CascadeType.PERSIST:级联新增(也称级联保存)。

- CascadeType.MERGE：级联合并。更新该实体时，与其有映射关系的实体也随之更新。
- CascadeType.REMOVE：级联删除。删除该实体时，与其有映射关系的实体也随之删除。
- CascadeType.REFRESH：级联刷新。该实体被操作前都会刷新，以保证数据合法性。
- CascadeType.ALL：包含以上 4 种级联操作。

（3）fetch 表示数据加载策略，默认值为 FetchType.EAGER。

- FetchType.LAZY：表示数据获取方式为懒加载，表示关系类在被访问时才加载。
- FetchType.EAGER：表示数据获取方式为急加载，表示关系类在主类加载的同时加载。

（4）optional 表示关联关系是否必须，当该值为 true 时，One 的一方可以为 null。

（5）mappedBy 指定映射关系由哪一方维护，一般使用在双向映射场景。在实际开发中，是采用单向关联还是双向关联，要看具体的业务需求，如果业务只需要在获取一方的实体时获取另一方的实体，而不是两边都能获取，那就采用单向关联。反之，如果需要在两边的实体中都能获取对方，就使用双向关联。如果关系是单向的就不需要，双向关系表中拥有关系的一方有建立、解除和更新与另一方关系的权限，而另一方只能被动管理。该属性被定义在关系的被拥有方，只有 @OneToOne、@OneToMany 和 @ManyToMany 才有 mapperBy 属性。

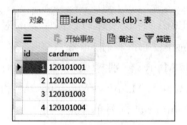

图 4-6　卡号表 idcard

（6）orphanRemoval 表示孤值删除，孤立数据和外键为 null 的数据将被删除。

下面通过示例讲解 Spring Data JPA 如何实现用户和卡号之间的一对一关系映射。

（1）以项目 springboot0405 为基础，创建项目 springboot0406，其中卡号表 idcard 和用户表 user 分别如图 4-6 和图 4-7 所示。

图 4-7　用户表 user

（2）创建持久化实体类。

User 类的具体代码如下：

```
@Entity
@Table(name = "user")
@Data
public class User implements Serializable {
    @Id
    private String id;
    private String password, username, address, email, phone;
```

```
    @OneToOne(optional = true)
    @JoinColumn(name = "idcard")
    private IdCard idCard;
}
```

@JoinColumn 与@Column 注解一样，用于注释表中的字段。它的属性与@Column 属性有很多相同之处，这里不再详细讲述。

@JoinColumn 与@Column 的区别是：@JoinColumn 注释的是保存表与表之间关系的字段，它要标注在实体属性上，而@Column 标注的是表中不包含表关系的字段。

与@Column 标记一样，name 属性用来标识表中所对应的字段的名称，如 user 表中存在的字段 idcard。若此时不设置 name 的值，则在默认情况下，name 的取值遵循以下规则：

  name = 关联表的名称 + "_" + 关联表主键的字段名

例如，在 User 实体中，如果不指定 name 的值，默认将对应 name=idcard_id。这是因为@JoinColumn 注释在实体 IdCard 属性上，实体 IdCard 对应的表名为 idcard，而表 idcard 的主键是 id，所以此时对应的默认字段名称为 idcard_id。

在默认情况下，关联的实体的主键一般是用来做外键的。如果此时不想主键作为外键，则需要设置 referencedColumnName 属性，标注所关联表中的字段名。

IdCard 类的具体代码如下：

```
@Entity
@Table(name = "idcard")
@Data
public class IdCard implements Serializable {
    @Id
    private String cardnum;
    private Integer id;
    @OneToOne(mappedBy = "idCard", optional = false)
    //@JsonIgnore 注解是类注解，作用是在 JSON 序列化时将 Java Bean 中的一些属性忽略掉
    @JsonIgnore
    private User user;
}
```

在上述代码中多了一个 mappedBy 方法，它表示当前所在表和 User 的关系是定义在 User 里面的 idCard 上的，即此表是一对一关系中的从表，关系是在 user 表里面维护的，这一点很重要。user 表是关系的维护者，有主导权，且有指向 IdCard 的外键。也可以让主导权存在于 IdCard 上，即让它产生一个指向 user 的外键，这也是可以的，但最好是让 User 来维护整个关系，这样更符合常规思维。

在 User 里面对 IdCard 的注释是 optional=true，也就是说一个人是可以没有卡号的，但是一个信用卡不可以没有人。因此在 IdCard 里面注释 user 时，optional 就取值为 false 了，这样可以防止一个空的信用卡记录写进数据库。

（3）创建数据访问层 UserRepository。

```
public interface UserRepository extends JpaRepository<User, String> {
    User findByIdCard_Cardnum(String cnum);
}
```

依照 Spring Data JPA 的命名规则，对于两个有关联关系的对象的查询，可以通过方法

名中加"_"下画线来标识,也可以通过 Spring-Data-JPA 命名规范查询,同时 Spring Data JPA 还支持用@Query 注解定义在数据访问层接口的方法上实现查询。

(4) 创建业务层 UserServiceImpl。

```
@Service
public class UserServiceImpl implements UserService {
    @Autowired
    private UserRepository dao;
    @Override
    public User findUserByCardnum(String cnum) {
        return dao.findByIdCard_Cardnum(cnum);
    }
}
```

(5) 创建控制器类 UserController。

```
@Controller
public class UserController {
    @Autowired
    private UserService service;

    @RequestMapping("/loginadmin")
    public String findAll(Model model) {
        Calendar calendar = Calendar.getInstance();
        SimpleDateFormat dateFormat = new SimpleDateFormat("yyyy-MM-dd");
        model.addAttribute("dateString", dateFormat.format(calendar.getTime()));
        return "/admin/findUser";
    }

    @ResponseBody
    @GetMapping("/findCnum/{cnum}")
    public User findUserByCardnum(@PathVariable String cnum, Model model) {
        System.out.println(cnum);
        User user = service.findUserByCardnum(cnum);
        return user;
    }
}
```

(6) 运行程序,效果如图 4-8 所示。

图 4-8 一对一查询效果图

2. @OneToMany 和 @ManyToOne

在实际生活中,订单和产品是一对多的关系,那么在 Spring Data JPA 中如何描述一对多的关系呢？下面将使用@OneToMany 和@ManyToOne 来表示一对多的双向关联。

(1) ManyToOne(多对一)单向：不产生中间表,但可以用@Joincolumn(name="")来指定生成外键的名字,且外键在多的一方表中产生。

(2) OneToMany(一对多)单向：会产生中间表,此时可以用@Joincolumn(name="")避免产生中间表,并且指定外键的名字。

(3) OneToMany,ManyToOne(两个注解同时使用)双向：如果不在@OneToMany 中添加 mappedBy 属性就会产生中间表,此时通常在@ManyToOne 的注解下再添加注解@Joincolumn(name=" ")来指定外键的名字。

说明：多的一方为关系维护端,关系维护端负责外键记录的更新,关系被维护端没有更新外键记录的权限。@OneToMany(mappedBy="一对多中,多对一的属性")出现 mappedBy 则为被维护端,默认为延迟加载。

下面通过示例讲解 Spring Data JPA 如何实现 Order 和 OrderItem 之间的一对多关系映射。

(1) 打开项目 springboot0406,其中订单表 orders 和订单详情表 orderitem 分别如图 4-9 和图 4-10 所示。

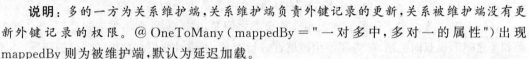

图 4-9　订单表 orders

图 4-10　订单详情表 orderitem

(2) 创建持久化实体类。

Order 类的具体代码如下。

```
@Entity
@Table(name = "orders")
@Data
public class Order implements Serializable {
    @Id
    @GeneratedValue(strategy = GenerationType.IDENTITY)
    @Column(name = "order_id")
```

```
    private Integer oid;
    @Column(name = "price_num")
    private Integer price;
    @Column(name = "order_date")
    private String odate;
    @OneToOne(optional = false)
    @JoinColumn(name = "user_id")
    private User user;
    @OneToMany(mappedBy = "order", cascade = CascadeType.REFRESH)
    @OrderBy("id asc ")
    private List<OrderItem> orderItems;
}
```

对上述代码的说明如下。

@OneToMany：指明 Order 与 OrderItem 关联关系为一对多关系。

mappedBy：定义类之间的双向关系。如果类之间是单向关系，则不需要提供定义。如果类和类之间形成双向关系，就需要使用该属性进行定义，否则可能引起数据一致性的问题。

cascade：CascadeType[]类型。该属性定义类和类之间的级联关系。定义的级联关系将被容器视为对当前类对象及其关联类对象采取相同的操作，而且这种关系是递归调用的。在实际业务中，通常会遇到以下情况：

① 用户和用户的收货地址是一对多关系，当用户被删除时，这个用户的所有收货地址也应该一并删除。

② 订单和订单中的商品也是一对多关系，但订单被删除时，订单所关联的商品肯定不能被删除。

此时只要配置正确的级联关系，就能达到想要的效果。常用的级联关系如下。

- CascadeType.REFRESH：级联刷新。当多个用户同时操作一个实体时，为了保证用户获取的数据是实时的，在用实体中的数据前需要调用 refresh()方法。
- CascadeType.REMOVE：级联删除。当调用 remove()方法删除 Order 实体时，会先级联删除 OrderItem 的相关数据。
- CascadeType.MERGE：级联更新。当调用 Merge()方法时，如果 Order 中的数据改变了，则会相应地更新 OrderItem 中的数据。
- CascadeType.ALL：包含以上所有级联属性。
- CascadeType.PERSIST：级联保存。当调用 Persist()方法时，会级联保存相应的数据。

需要注意的是，CascadeType.ALL 要谨慎使用，为了达到数据同步，很多人喜欢用 CascadeType.ALL 来实现，但上面订单和商品的例子就不适用。

fetch：可选择项包括 FetchType.EAGER 和 FetchType.LAZY。前者表示关系类（本例是 OrderItem 类）在主类（本例是 Order 类）加载的同时加载，后者表示关系类在被访问时才加载，默认值是 FetchType.LAZY。

@OrderBy("id asc")：指明加载 OrderItem 时按 id 的升序排序。

OrderItem 类的具体代码如下。

```
@Entity
@Table(name = "orderitem")
@Data
public class OrderItem implements Serializable {
```

```
    @Id
    @GeneratedValue(strategy = GenerationType.IDENTITY)
    private Integer id;
    private String isbn;
    @Column(name = "book_num")
    private Integer bnum;
    @ManyToOne(optional = false, cascade = {CascadeType.MERGE,CascadeType.REFRESH})
    @JoinColumn(name = "order_id")
    @JsonIgnore
    private Order order;
}
```

@ManyToOne：指明 OrderItem 和 Order 之间为多对一关系，多个 OrderItem 实例关联的都是同一个 Order 对象。其中的属性和@OneToMany 基本一样，但@ManyToOne 注释的 fetch 属性默认值是 FetchType.EAGER，optional 属性的默认值是 true。例如，某项订单(Order)中没有订单项(OrderItem)，如果 optional 属性设置为 false，获取该项订单(Order)时，得到的结果为 null。如果 optional 属性设置为 true，仍然可以获取该项订单，但订单中指向订单项的属性为 null。

实际上在解释 Order 与 OrderItem 的关系时，optional 属性指定了 optional=false 连接关系为 inner join，optional=true 连接关系为 left join。

@JoinColumn：指明被维护端(OrderItem)的外键字段为 order_id。

@JoinColumn 标签和@OneToMany 中 mappedBy 属性的作用相同，都是用于指定关联键的，区别在于 mappedBy 指定的是实体类中的属性，而@JoinColumn 指定的是数据表里面的字段。

(3) 创建数据访问层 OrderRepository。

```
public interface OrderRepository extends JpaRepository<Order, Integer> {
    Order findByOid(Integer oid);
}
```

(4) 创建业务层 OrderServiceImpl。

```
@Service
public class OrderServiceImpl implements OrderService {
    @Autowired
    private OrderRepository dao;
    @Override
    public Order findOrderByOid(Integer oid) {
        return dao.findByOid(oid);
    }
}
```

(5) 创建控制器类 UserController。

```
@Controller
public class OrderController {
    @Autowired
    private OrderService service;

    @RequestMapping("/loginOrder")
    public String findAll(Model model) {
        Calendar calendar = Calendar.getInstance();
        SimpleDateFormat dateFormat = new SimpleDateFormat("yyyy-MM-dd");
```

```
            model.addAttribute("dateString", dateFormat.format(calendar.getTime()));
            return "/admin/findOrder";
        }

        @ResponseBody
        @GetMapping("/findOrder/{oid}")
        public Order findOrderByOid(@PathVariable Integer oid, Model model) {
            Order order = service.findOrderByOid(oid);
            return order;
        }
    }
```

（6）运行程序，效果如图 4-11 所示。

图 4-11　一对多运行效果图

视频讲解

3. @ManyToMany

在实际生活中，用户和权限是多对多的关系。一个用户可以有多个权限，一个权限也可以被很多用户拥有。

在 Spring Data JPA 中，使用@ManyToMany 注解多对多的映射关系，由一个关联表来维护。关联表的表名默认为主表名＋下画线＋从表名（主表是指关系维护端对应的表，从表是指关系被维护端对应的表）。关联表只有两个外键字段，分别指向主表 ID 和从表 ID。字段的名称默认为主表名＋下画线＋主表中的主键列名，从表名＋下画线＋从表中的主键列名。需要注意的是，多对多关系中一般不设置级联保存、级联删除、级联更新等操作。

下面通过示例讲解 Spring Data JPA 如何实现 User 和 Role 之间的多对多关系映射。

（1）打开项目 springboot0406，其中权限表 role 和用户权限明细表 user_role 分别如图 4-12 和图 4-13 所示。

图 4-12　权限表 role

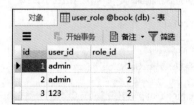

图 4-13　用户权限明细表 user_role

（2）创建持久化实体类。

User 类的具体代码如下：

```java
@Entity
@Table(name = "user")
@Data
public class User implements Serializable {
    @Id
    private String id;
    private String password, username, address, email, phone;
    @OneToOne(optional = true)
    @JoinColumn(name = "idcard")
    private IdCard idCard;

    @ManyToMany
    @JoinTable(name = "user_role", joinColumns = @JoinColumn(name = "user_id"),
inverseJoinColumns = @JoinColumn(name = "role_id"))
    private List<Role> roles;
}
```

@ManyToMany：表示 User 是多对多关系的一端。

@JoinTable：描述多对多关系的数据表关系。name 属性指定中间表名称，joinColumns 定义中间表与 user 表的外键关系。在上面的代码中，中间表 user_role 的 user_id 列是 user 表的主键列对应的外键列，inverseJoinColumns 属性定义了中间表与另外一端（Role）的外键关系。

Role 类的具体代码如下：

```java
@Data
@Table(name = "role")
@Entity
public class Role implements Serializable{
    @Id
    @GeneratedValue(strategy = GenerationType.IDENTITY)
    private Integer id;
    private String rolename;

    @ManyToMany(mappedBy = "roles")
    private List<User> user;
}
```

拥有 mappedBy 注解的实体类为关系被维护端，另外的实体类为关系维护端。顾名思义，关系的维护端对关系（在多对多关系中为中间关联表）做 CRUD 操作，关系的被维护端没有该操作且不能维护关系。

当关系维护端被删除时，如果中间表存在记录的关联信息，则会删除该关联信息；当关系被维护端被删除时，如果中间表存在记录的关联信息，则会删除失败。

（3）创建数据访问层 UserRepository。

```java
public interface UserRepository extends JpaRepository<User, String> {
    User findByIdCard_Cardnum(String cnum);
    User findByUsername(String username);
}
```

（4）创建业务层 UserServiceImpl。

```
@Service
public class UserServiceImpl implements UserService {
    @Autowired
    private UserRepository dao;
    @Override
    public User findByUsername(String username) {
        return dao.findByUsername(username);
    }
}
```

（5）创建控制器类 UserController。

```
@Controller
public class OrderController {
    @Autowired
    private OrderService service;

    @RequestMapping("/loginRole")
    public String loginRole( ) {
        return "/admin/findUserRole";
    }
    @ResponseBody
    @GetMapping("/findUserRole/{username}")
    public User findUserRole(@PathVariable String username, Model model) {
        User user = service.findByUsername( username);
        return user;
    }
}
```

（6）运行程序，效果如图 4-14 所示。

图 4-14　多对多运行效果图

4.3.4　@Query 和 @Modifying 注解

1. @Query 注解

@Query 注解查询适用于无法通过关键字查询数据得到结果的情况。这种查询可以摆脱像关键字查询那样的约束，将查询直接在相应的接口方法中声明，结构更为清晰，这是 Spring Data 的特有实现。

@Query 注解常用属性如下。

- value：取值。要么使用原生 SQL，要么使用 Java 持久化查询语言（Java Persistence Query Language，JPQL）。
- nativeQuery=true：表示是本地查询。即使用原生的 SQL 语句，直接查询数据表名，而不是通过实体类对象进行查询数据库的操作。

```
@Query(value = "select * from book b where b.name = ?1", nativeQuery = true)
List<Book> findByName(String name);
```

book 为数据库中真实存在的数据表，name 为 book 表中的字段。当不设置 nativeQuery=true 时，value 取值为 JPQL。在 JPQL 的语法中，表名的位置对应 Entity 的名称，字段对应 Entity 的属性。两种方式作用一样，只是方法不同。

```
@Query(value = "select b from Book b where b.name = ?1")
List<Book> findByName(String name);
```

Book 为 Entity 对象名，如果将 Entity 对象的 name 设置为"Book"，则 Entity 对象名就是 name 后的名字；如果没有设置 name，则默认使用类名。where 后为对象别名.属性名，这里的属性名不是@Column(name="")的 name 的名字，而是当前 Entity 对象的 property，即 private String name 中的 name。

下面介绍@Query 注解使用的几种情况。

(1) 索引参数。

索引值从 1 开始，查询中的"?X"个数需要与方法定义的参数个数一致，并且顺序也要一致。

```
@Query("SELECT b FROM Book b WHERE b.title = ?1 Or b.author = ?2")
List<Book> findBooksByTitleAuthor(String title, String author);
```

注意：上面代码中的?1 和?2 表示参数的占位符，需要与方法中所传递的参数顺序一致。

(2) 命名参数(推荐使用此方式)。

可以定义好参数名，赋值时使用@Param("参数名")，而不用处理顺序。

```
@Query("SELECT b FROM Book b WHERE b.title = :title Or b.author = :writer")
List<Book> findBooksByTitleAuthor(@Param("writer") String author,@Param("title") String title);
```

注意：上面代码中的:title 和:writer 表示为参数命名，方法中所传递的参数使用@Param 注解标识命名参数，这种方式不用处理参数的顺序。

(3) 含有 LIKE 关键字的查询。

方式 1：可以在占位符上添加"%"，这样在查询方法中就不用添加"%"。

```
@Query("SELECT b FROM Book b WHERE b.title like %?1% Or b.author like %?2%")
List<Book> findBooksByTitleAuthor(String title, String author);
```

方式 2：不在占位符上添加"%"，这样就必须在查询方法的参数上添加"%"。

```
@Query("SELECT b FROM Book b WHERE b.title like ?1 Or b.author like ?2")
List<Book> findBooksByTitleAuthor(String title, String author);
```

方式 3：在命名参数上添加"%"。

```
@Query("SELECT b FROM Book b WHERE b.title like % :title% Or b.author like % :writer%")
```

```
List<Book> findBooksByTitleAuthor(@Param("writer") String author,@Param("title") String title);
```

（4）对象参数。

```
@Query("SELECT b FROM Book b WHERE b.title = :#{#book.title}")
List<Book> findBooksByTitle(@Param("book") Book book);
```

（5）Spring 表达式。

在使用@Entity 注解后，#{#entityName}会取@Entity()的值，默认是类名小写。如果已经设置为@Entity(name = "bookinfo")，则取出的值就是 bookinfo。

```
@Query("SELECT b FROM #{#entityName} b WHERE b.title like %:t%")
List<Book> findBooksByTitleAuthor(@Param("t") String title);
```

（6）@Query 返回自定义字段。

HQL 和原生 SQL 都可以实现。HQL(Hibernate Query Language，Hibernate 查询语言)是一种面向对象的查询语言，类似于 SQL，但并不对表和列进行操作，而是面向对象和它们的属性。

```
@Query("SELECT b.ISBN,b.title,b.author FROM Book b WHERE b.title = :#{#book.title}")
List<String[]> findBooksByTitleAuthor(@Param("book") Book book);
```

（7）@Query 返回自定义对象。

使用的是 HQL 语法，不支持 SQL 原生，from 后面都是用的对象。

```
@Query("SELECT new map (b.ISBN,b.title,b.author) FROM Book b WHERE b.title = :#{#book.title}")
List<Map<String,String>> findBooksByTitleAuthor(@Param("book") Book book);
```

（8）使用原生 SQL 进行查询。

设置 nativeQuery=true 即可使用原生 SQL 进行查询。

```
@Query(value = "SELECT o.*,u.* FROM user AS u,orders AS o WHERE u.username LIKE %?1% and u.id=o.USER_ID",nativeQuery = true)
    public List<Order> findAllOrdersByUsername(String username);
```

2. @Modifying 注解

在@Query 注解中编写 JPQL 实现 DELETE 和 UPDATE 操作时，必须加上@modifying 注解和@Transactional 注解，以通知 Spring Data 这是一个 DELETE 或 UPDATE 操作。在执行 UPDATE 或 DELETE 操作时，需要在 Service 层的方法上添加事务操作。

Spring Data 提供了默认的事务处理方式，即所有的查询均声明为只读事务。对于自定义的方法，如需改变 Spring Data 提供的事务默认方式，可以在方法上注解 @Transactional 声明。

在进行多个 Repository 操作时，也应该在同一个事务中处理，按照分层架构的思想，这部分属于业务逻辑层。因此，需要在 Service 层实现对多个 Repository 的调用，并在相应的方法上声明事务。

此外需要注意的是，JPQL 不支持 INSERT 操作。

下面通过示例讲解 Spring Data JPA 如何实现@Query 和@Modifying 注解的使用。

（1）打开项目 springboot0406。

（2）修改数据访问层 OrderRepository。

```java
public interface OrderRepository extends JpaRepository<Order, Integer> {
    Order findByOid(Integer oid);
    @Query("SELECT o FROM Order o where o.user.username like %?1%")
    public List<Order> findAllOrdersByUsername(String username);

    @Query("delete from Order o where o.oid = ?1")
    @Modifying
    public int deleteOrderByOid(Integer oid);
}
```

(3) 修改业务层 OrderServiceImpl。

```java
@Service
public class OrderServiceImpl implements OrderService {
    @Autowired
    private OrderRepository dao;
    @Override
    public Order findOrderByOid(Integer oid) {
        return dao.findByOid(oid);
    }
    @Override
    public List<Order> findAllOrdersByUsername(String username) {
        return dao.findAllOrdersByUsername(username);
    }
    @Override
    @Transactional
    public boolean deleteOrderByOid(Integer oid) {
      int result = dao.deleteOrderByOid(oid);
      return result > 0?true:false;
     }
}
```

(4) 修改控制器类 OrderController。

```java
@Controller
public class OrderController{
    @Autowired
    private OrderService service;

    @RequestMapping("/findOrders")
    public String toFindOrders() {
     return "admin/findAllOrders";
    }
}

@ResponseBody
@GetMapping("/findAllOrders/{username}")
public String findAllOrders(@PathVariable String username) {
  List<Order> orders = service.findAllOrdersByUsername(username);
  System.out.println(orders.size());
  JSONArray jsonArray = new JSONArray();
  JSONObject jsonObject = new JSONObject();
  for (Order order : orders) {
    JSONObject object = new JSONObject();
    object.put("oid", order.getOid());
    object.put("username", order.getUser().getUsername());
    object.put("price", order.getPrice());
```

```
            object.put("odate",order.getOdate());
            jsonArray.add(object);
        }
        jsonObject.put("orders", jsonArray);
        return jsonObject.toJSONString();
    }
    @RequestMapping("/deleteOrder/{oid}")
    public String toDeleteOrder(@PathVariable String oid) {
        service.deleteOrderByOid(new Integer(oid));
        return "redirect:/findOrders";
    }
}
```

（5）运行程序，效果如图4-15所示。

图4-15　模糊查询效果图

4.3.5　排序和分页查询

如果一次性加载成千上万的列表数据，在网页上显示将十分地耗时，用户体验不好。所以处理较大数据查询结果显示时，分页查询是必不可少的。分页查询必然伴随着一定的排序规则，否则分页数据的状态很难控制，导致用户可能在不同的网页看到同一条数据。那么，如何使用Spring Data JPA进行分页与排序呢？

下面通过示例讲解Spring Data JPA实现排序和分页查询的方法。

（1）打开项目springboot0406。

（2）修改数据访问层OrderRepository。

```
public interface OrderRepository extends JpaRepository<Order, Integer> {
    //查询order表的所有数据，传入Pageable分页参数，不需要自己写SQL
    Page<Order> findAll(Pageable pageable);
    //根据关联的User属性的username字段模糊查询order表数据，传入Pageable分页参数，不需要自己写SQL
    Page<Order> findByUser_UsernameContaining(String username, Pageable page);
    //根据odate字段和price字段查询order表数据，传入Pageable分页参数，不需要自己写SQL
    Slice<Order> findByOdateOrAndPrice(String odate, Integer price, Pageable pageable);
}
```

Pageable是Spring定义的接口，用于分页参数的传递。首先将OrderRepository注入需要进行持久层操作的类里面，通常是一个@Service注解的类，然后在服务方法内进行分页操作。下面所示代码为查询第 1 页（从 0 开始）的数据，每页 10 条数据。

```
Pageable pageable = PageRequest.of(0, 10);          //第 1 页
//Pageable pageable = PageRequest.of(1, 10);        //第 2 页
//Pageable pageable = PageRequest.of(2, 10);        //第 3 页
//数据库操作获取查询结果
Page<Article> articlePage = articleRepository.findAll(pageable);
//将查询结果转换为 List
List<Article> articleList = articlePage.getContent();
```

（3）修改业务服务层 OrderServiceImpl。

```
@Service
public class OrderServiceImpl implements OrderService {
    @Autowired
    private OrderRepository dao;
    @Override
    public MyOrderPage findByUsernamePage(String username, Integer pageNum) {
        Sort sort = Sort.by(Sort.Direction.ASC,"price");
        Pageable page = PageRequest.of(pageNum - 1, 2, sort);
        Page<Order> pageData = dao.findByUser_UsernameContaining(username, page);
        MyOrderPage myOrderPage = new MyOrderPage();
        myOrderPage.setOrders(pageData.getContent());
        myOrderPage.setPage(pageNum);
        myOrderPage.setTotalCount(pageData.getTotalElements());
        myOrderPage.setTotalPage(pageData.getTotalPages());
        return myOrderPage;
    }
}
```

Spring Data JPA 提供了一个 Sort 对象，用于提供一种排序机制，其排序方式如下。

```
dao.findAll(Sort.by("odate"));
dao.findAll(Sort.by("author").ascending().and(Sort.by("odate").descending()));
```

第一个 findAll 方法是按照 odate 的升序进行排序。第二个 findAll 方法是按照 author 的升序进行排序，再按照 odate 的降序进行排序。

分页和排序同时进行：

```
Pageable pageable = PageRequest.of(0, 10,Sort.by("odate"));
```

在 OrderRepository 中，一个方法返回 Slice，而另一个方法返回 Page。它们都是 Spring Data JPA 的数据响应接口，都用于保存和返回数据，其中 Page 是 Slice 的子接口。

Slice 的一些重要方法如下。

List<T> getContent()：获取切片的内容。

Pageable getPageable()：获取当前切片的分页信息。

boolean hasContent()：显示是否有查询结果。

boolean isFirst()：显示是否为第一个切片。

boolean isLast()：显示是否为最后一个切片。

Pageable nextPageable()：显示下一个切片的分页信息。

Pageable previousPageable()：显示上一个切片的分页信息。

Page 是 Slice 的子接口，其重要方法如下。

int getTotalPages()：获取总页数。

long getTotalElements()：获取总数据条数。

那么，什么时候使用 Slice 和 Page 呢？通过这两个接口的函数定义可以看出，Slice 只关心是否存在下一个分片（分页），不会去数据库计算总条数和总页数，所以比较适合大数据量列表的鼠标或手指滑屏操作。Page 比较适合传统应用中的 Table 开发，需要知道总页数和总条数。

在前端展示一个分页列表，不仅需要数据，而且还需要一些分页信息。例如，当前第几页，每页多少条，总共多少页，总共多少条等。这些信息都可以在 pageData 对象里面进行获取，在此创建一个存储当前分页信息的类是必不可少的。

（4）创建分页信息类 MyOrderPage。

```java
@Data
public class MyOrderPage {
    private List<Order> orders;          //当前页面上的数据
    private Integer page;                //当前页码
    private Long totalCount;             //共多少条记录
    private Integer totalPage;           //共多少页
}
```

（5）修改控制器类 OrderController。

```java
@Controller
public class OrderController {
    @Autowired
    private OrderService service;
    @ResponseBody
    @GetMapping("/findAllOrdersPage/{username}/{page}")
    public String findAllOrdersPage(@PathVariable String username, @PathVariable Integer page) {
        MyOrderPage myOrderPage = service.findByUsernamePage(username, page);
        JSONArray jsonArray = new JSONArray();
        JSONObject jsonObject = new JSONObject();
        for (Order order : myOrderPage.getOrders()) {
            JSONObject object = new JSONObject();
            object.put("oid", order.getOid());
            object.put("username", order.getUser().getUsername());
            object.put("price", order.getPrice());
            object.put("odate", order.getOdate());
            jsonArray.add(object);
        }
        jsonObject.put("orders", jsonArray);
        jsonObject.put("page", page);
        jsonObject.put("totalPage", myOrderPage.getTotlePage());
        jsonObject.put("totalCount", myOrderPage.getTotleCount());
        return jsonObject.toJSONString();
    }
}
```

（6）创建 View 视图页面。

```
<script type="text/javascript">
```

```javascript
function findformPage(page) {
    var title = document.getElementById("tbody-title");
    var username = $("#username").val();
    $.ajax({
        url: "findAllOrdersPage/" + username + "/" + page,
        dataType: "json",
        type: "get",
        success: function (data) {
            var titles = "";
            for (var i = 0; i < data.orders.length; i++) {
                var item = data.orders[i];
                titles += "<tr><td><div align='center'><span>" + (i + 1) + "</span></div></td>" + "<td height='32'><div align='center'><span>" + item.username + "</span></div></td>" + "<td><div align='center'><span>" + item.price + "</span></div></td>" + "<td><div align='center'><span>" + item.odate + "</span></div></td>" + "<td><div align='center'><span><a href='deleteOrder/" + item.oid + "'>删除</a></span></div></td></tr>";
            }
            titles += "<tr><td colspan='5' align='right'>";
            if (data.page != 1) {
                titles += "<a href='javascript:findformPage(" + (data.page - 1) + ")'>上一页</a> ";
            }
            titles += "第" + (data.page) + "<span>页</span> ";
            titles += "共" + (data.totalPage) + "<span>页</span> ";
            titles += "共" + (data.totalCount) + "<span>条</span> ";
            if (data.page != data.totalPage) {
                titles += "<a href='javascript:findformPage(" + (data.page + 1) + ")'>下一页</a>  ";
            }
            titles += "</td></tr>";
            title.innerHTML = titles;
        }
    })
}
</script>
```

(7) 运行程序,效果如图 4-16 所示。

图 4-16　分页效果图

4.4 数据缓存 Cache

一个程序的瓶颈在于数据库,且内存的速度远远快于硬盘的速度。当需要重复地获取相同的数据时,重复地请求数据库或远程服务会导致大量的时间耗费在数据库查询或远程方法调用上,进而导致程序性能的恶化,这便是数据缓存要解决的问题。

视频讲解

Spring 定义了 org.springframework.cache.CacheManager 和 org.springframework.cache.Cache 接口用来统一不同的缓存技术,其中 CacheManager 是 Spring 提供的各种缓存技术的抽象接口,Cache 接口包含缓存的各种操作(如增加、删除和获取缓存等,一般不直接使用此接口)。CacheManager 接口描述如表 4-3 所示。

表 4-3 CacheManager 接口描述

CacheManager	描述
SimpleCacheManager	使用简单的 Collection 存储缓存,主要用来测试用途
ConcurrentMapCacheManager	使用 ConcurrentMap 存储缓存
NoOpCacheManager	仅测试用途,不会实际存储缓存
EhCacheCacheManager	使用 EhCache 作为缓存技术
GuavaCacheCacheManager	使用 Google Guava 的 GuavaCache 作为缓存技术
HazelcastCacheManager	使用 Hazelcast 作为缓存技术
JCacheCacheManager	使用支持 JCache(JSR-107)标准的技术作为缓存技术,如 Apache Commons JCS
RedisCacheManager	使用 Redis 作为缓存技术

在使用任意一个实现的 CacheManager 时,需要先注册实现的 CacheManager 的 Bean,例如:

```
@Bean
public EhCacheCacheManager cacheManager(CacheManager ehCacheCacheManager) {
    return new EhCacheCacheManager(ehCacheCacheManager);
}
```

一旦配置好 Spring 缓存支持,就可以在 Spring 容器管理的 Bean 中使用缓存注解,一般情况下都是在业务层使用这些注解。

1. @Cacheable

@Cacheable 可以标记在一个方法上,也可以标记在一个类上。当标记在一个方法上时,表示该方法是支持缓存的;当标记在一个类上时,表示该类所有的方法都是支持缓存的。对于一个支持缓存的方法,Spring 会在其被调用后将其返回值进行缓存,以保证下次利用同样的参数执行该方法时可以直接从缓存中获取结果,而不需要再次执行该方法。Spring 在缓存方法的返回值时是以键值对进行缓存的,值就是方法的返回结果,键则支持默认策略和自定义策略两种策略。

@Cacheable 可以指定 3 个属性:value、key 和 condition。

value 属性指定 Cache 的名称,该属性是必须指定的。它表示当前方法的返回值缓存在哪个 Cache 上,并对应 Cache 的名称。它可以指定一个或多个 Cache,当需要指定多个 Cache 时是一个数组。

```java
@Cacheable("cache1")//Cache 是发生在 cache1 上的
public User find(Integer id) {
    return null;
}
@Cacheable({"cache1", "cache2"})//Cache 是发生在 cache1 和 cache2 上的
public User find(Integer id) {
    return null;
}
```

key 属性用来指定 Spring 缓存方法在返回结果时对应的 key。当没有指定该属性时，Spring 将使用默认策略生成 key，也可以通过 Spring 的 EL 表达式指定 key。该属性支持 Spring EL 表达式，可以使用方法参数及它们对应的属性，使用方法参数时可以直接使用"♯参数名"或者"♯p 参数 index"。下面是几个使用参数作为 key 的示例。

```java
//表示传入时的参数
@Cacheable(value = "users", key = "♯id")
public User find(Integer id) {
    return null;
}
//表示第一个参数
@Cacheable(value = "users", key = "♯p0")
public User find(Integer id) {
    return null;
}
//表示 User 中的 id 值
@Cacheable(value = "users", key = "♯user.id")
public User find(User user) {
    return null;
}
//表示第一个参数里的 id 属性值
@Cacheable(value = "users", key = "♯p0.id")
public User find(User user) {
    return null;
}
```

有时可能并不希望缓存一个方法所有的返回结果，通过 condition 属性可以实现这一功能。condition 属性默认为空，表示将缓存所有的调用情形。该属性值是通过 Spring EL 表达式指定的，当为 true 时表示进行缓存处理；当为 false 时表示不进行缓存处理，即每次调用该方法时都会执行一次。例如，下面的示例表示只有当 user 的 id 为偶数时才会进行缓存。

```java
//根据条件判断是否缓存
@Cacheable(value = {"users"}, key = "♯user.id", condition = "♯user.id％2 == 0")
public User find(User user) {
    System.out.println("find user by user " + user);
    return user;
}
```

2. @CachePut

@CachePut 表明 Spring 应该将方法的返回值放到缓存中。在方法调用前并不会检查缓存，方法始终都会被调用。@CachePut 与 @Cacheable 的属性一致。

3. @CacheEvict

@CacheEvict 标注在需要清除缓存元素的方法或类上。当标记在一个类上时，表示其中所有方法的执行都会触发缓存的清除操作。@CacheEvict 可以指定的属性有 value、key、condition、allEntries 和 beforeInvocation，其中 value、key 和 condition 的语义与@Cacheable 对应的属性类似。下面介绍新出现的两个属性：allEntries 和 beforeInvocation。

allEntries 是 boolean 类型，表示是否需要清除缓存中的所有元素。该属性值默认为 false，表示不需要。当指定 allEntries 为 true 时，Spring Cache 将忽略指定的 key。有时需要 Cache 一次性清除所有的元素，这比一个一个清除元素更有效率。

```
@CacheEvict(value = "users", allEntries = true)
public void delete(Integer id) {
    System.out.println("delete user by id: " + id);
}
```

清除操作默认是在对应方法成功执行后触发的，即方法如果因为抛出异常而未能成功返回时也不会触发清除操作。使用 beforeInvocation 可以改变触发清除操作的时间，当指定该属性值为 true 时，Spring 会在调用该方法前清除缓存中的指定元素。

```
@CacheEvict(value = "users", beforeInvocation = true)
public void delete(Integer id) {
    System.out.println("delete user by id: " + id);
}
```

4. @Caching

@Caching 组合多个注解策略在一个方法上，多用于处理复杂规则的数据缓存。该注解包含 cacheable、put 和 evict 这 3 个属性，其作用等用于@Cacheable、@CachePut 和 @CacheEvict。

```
@Caching(
    cacheable = {
        @Cacheable(cacheNames = {"user"},key = "#name") //(1)根据 name 查询 user
    },
    put = {
        @CachePut(cacheNames = {"user"},key = "#result.id") //(2) 根据 id 查询 user,以另一种 key 将查询出的结果放到缓存中
    }
)
@Override public User selectByName(String name) {
    return userMapper.selectByName(name);
}
```

这种映射规则将数据库返回结果{"id":1,"name":"张三","pwd":"123456"}放入 user 缓存块中，其他接口在获取数据时只需要指定以上两种 key 中的一个便可获取缓存中指定的数据。

下面通过示例讲解 Spring Boot 实现缓存的方法。

(1) 打开项目 springboot0406，在 Maven 中添加依赖以便启用 Spring Cache。

```
<dependency>
    <groupId>org.springframework.boot</groupId>
    <artifactId>spring-boot-starter-cache</artifactId>
</dependency>
```

（2）在启动类加上@EnableCaching注解即可开启使用缓存。

```
@SpringBootApplication
@EnableCaching    //开启缓存注解
public class Springboot0406Application{
    public static void main(String[] args) {
        SpringApplication.run(Springboot0406Application.class, args);
    }
}
```

（3）编写BookServiceImpl。

```
@CacheConfig(cacheNames = "book")
@Service
public class BookServiceImpl implements BookService {
    @Autowired
    private BookRepository dao;

    @Cacheable(key = "#isbn")
    @Override
    public Book findBookByIsbn(String isbn) {
        System.out.println("为key = " + isbn + "数据做了Cacheable 缓存");
        return dao.findByISBN(isbn);
    }

    @CachePut(key = "#book.ISBN")
    @Override
    public Book addBook(Book book) {
        System.out.println("为key = " + book.getISBN() + "数据做了CachePut 添加缓存");
        return dao.save(book);
    }

    @CachePut(key = "#book.ISBN")
    @Override
    public Book updateShop(Book book) {
        System.out.println("为key = " + book.getISBN() + "数据做了CachePut 修改缓存");
        return dao.save(book);
    }

    @CacheEvict(key = "#isbn")
    @Override
    public void delShop(String isbn) {
        System.out.println("删除 key = " + isbn + "数据做了CacheEvict 缓存");
        dao.deleteById(isbn);
    }

}
```

（4）编写BookController。

```
@Controller
public class BookController {
    @Autowired
    private BookService service;

    @ResponseBody
```

```java
        @RequestMapping("/find/{isbn}")
        public Book find(@PathVariable String isbn)
        {
            return service.findBookByIsbn(isbn);
        }

        @ResponseBody
        @RequestMapping("/add")
        public String add(Book book)
        {
            service.addBook(book);
            return "add -- ok";
        }

        @ResponseBody
        @RequestMapping("/update")
        public String update(Book book)
        {
            service.updateBook(book);
            return "update - ok";
        }

        @ResponseBody
        @RequestMapping("/delete/{isbn}")
        public String delete(@PathVariable String isbn)
        {
            service.delBook(isbn);
            return "delete - ok";
        }
}
```

(5) 运行测试@Cacheable。

启动应用程序的主类后,第一次访问 http://localhost:8080/find/1202255895 将调用方法查询数据库,并将查询到的数据存储到缓存 book 中,此时控制台输出结果如图 4-17 所示,页面数据如图 4-18 所示。

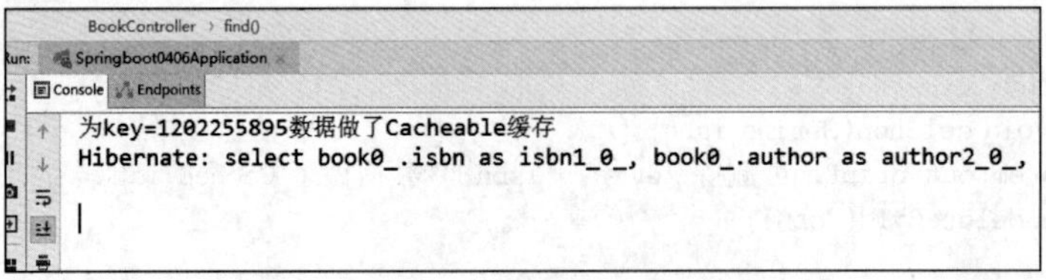

图 4-17 第一次访问查询控制台输出结果

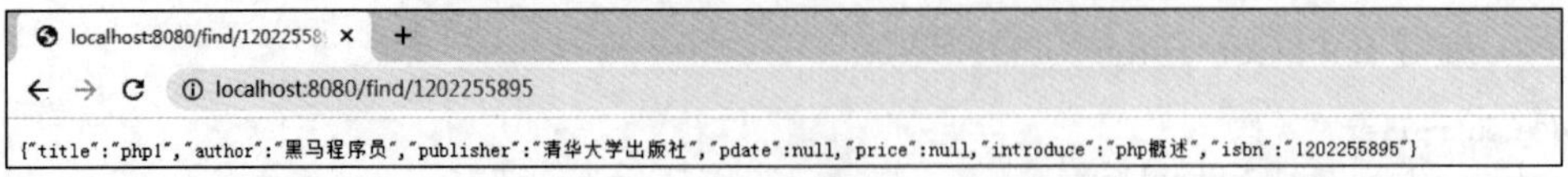

图 4-18 第一次访问查询页面数据

再次访问 http://localhost:8080/find/1202255895,此时控制台没有输出任何内容,这表明没有调用查询方法,页面数据直接从数据缓存中获得。

(6) 运行测试@CachePut。

重启应用程序的主类,访问 http://localhost:8080/add?ISBN=1202255897&title=网络技术&author=张三&publisher=清华大学出版社&introduce=网络概述,此时控制台输出结果如图 4-19 所示,页面数据如图 4-20 所示。

图 4-19 测试@CachePut 控制台输出结果

图 4-20 测试@CachePut 页面数据

再次访问 http://localhost:8080/find/1202255897,此时控制台无输出,页面从缓存直接获得数据。

(7) 运行测试@CacheEvict。

重启应用程序的主类。首先访问 http://localhost:8080/find/1202255897,为 key 为 1202255897 的数据做缓存,再次访问 http://localhost:8080/find/1202255897,确认数据已从缓存中获取。然后访问 http://localhost:8080/delete/1202255897,从缓存 book 中删除 key 为 1202255897 的数据,此时控制台输出结果如图 4-21 所示。

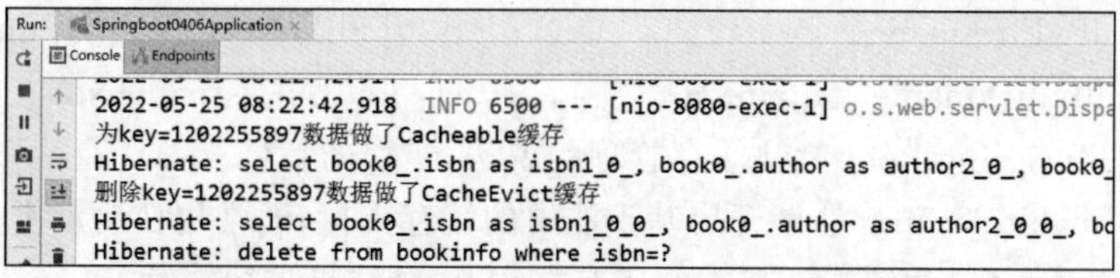

图 4-21 测试@CacheEvict 删除缓存数据

最后再次访问 http://localhost:8080/find/1202255897,此时重新做了缓存,控制台输出结果如图 4-22 所示。

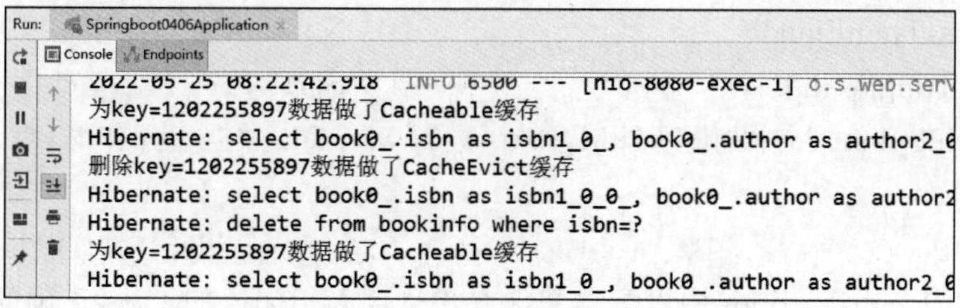

图 4-22 测试@CacheEvict 重做缓存数据

本章小结

本章是本书的重点章节,重点讲解了 Spring Data JPA、Spring Boot 整合 MyBatis 和数据缓存 Cache。本章选取了 Spring Boot 中几个主要的整合技术进行讲解,读者在学习过程中务必仔细查看并动手实践,同时要深刻体会 Spring Boot 与其他技术的整合思路。

在线测试

习题

一、单选题

1. 以下关于 Spring Boot 缓存注解的说法,正确的是( )。
   A. @Cacheable 注解是由 Spring Boot 框架提供的,可以作用于类或方法上
   B. @CachePut 注解的执行顺序是先进行方法调用,然后将方法结果更新到缓存中
   C. @CacheEvict 注解的执行顺序可以是先进行缓存清除,再进行方法调用
   D. @Caching 注解用于针对复杂规则的数据缓存管理,其内部包含有 Cacheable、put 和 evict 这 3 个属性

2. 以下关于 Spring Data JPA 映射实体类注解的说法,正确的是( )。
   A. @Entity 标注在类上,表示与数据表具有映射关系的实体类,必须使用 name 属性指定具体映射的表名
   B. @Id 必须标注在类属性上,表示某一个属性对应表中的主键
   C. @Column 标注在属性上,必须配合 name 属性表示类属性对应的表字段名
   D. @Transient 表示该属性并非一个到数据库表的字段的映射

3. 以下使用 JPA 中支持的方法名关键字构造 Repository 接口方法名,错误的是( )。
   A. findByFirstname   B. findByAgeLessAndEqual
   C. findByFirstnameContaining   D. findByLastnameNot

4. 以下关于 Spring Boot 默认缓存管理的说法,错误的是( )。
   A. @EnableCaching 注解开启基于注解的缓存支持,通常用在启动类上
   B. @Cacheable 注解标注在类的所有方法上,对结果进行缓存
   C. 在缓存管理中,每执行一次查询操作,本质是执行同样的 SQL 语句
   D. 在 Spring Boot 默认缓存管理中,不必配置 spring.jpa.show-sql=true

5. 以下 JPA 中支持的方法名关键字,错误的是( )。
   A. LessThanEqual   B. NotNull
   C. StartingWith   D. SortBy

6. 在 Spring Boot 2.1.3 中整合 MyBatis 进行 MySQL 数据库操作时,默认使用的数据源为( )。
   A. C3P0   B. Druid   C. tomcat.jdbc   D. hikari

7. 以下关于 Spring Boot 中以配置文件方式整合 MyBatis 的说法,正确的是( )。
   A. 以 XML 映射文件方式整合 MyBatis 时,可以不用添加 @Mapper 或 @MapperScan 注解

  B. 无论 XML 映射文件是否和 mapper 接口文件同目录,都必须在 Spring Boot 配置文件中指定 XML 映射文件位置

  C. 必须在 Spring Boot 配置文件中进行类的别名配置

  D. 以 XML 映射文件方式整合 MyBatis 时,需要在 Spring Boot 配置文件中配置驼峰命名映射

二、判断题

1. Spring Boot 整合 MyBatis 使用注解方式进行数据查询,可以将数据库数据完整映射到对应实体类字段上。（　　）

2. 将@CacheEvict 注解的 beforeInvocation 属性设置为 true,这会存在一定的弊端。
（　　）

3. 在 Spring Boot 中使用 XML 映射文件方式整合 MyBatis 时,必须在配置文件中配置类别名。（　　）

4. Spring Data JPA 映射实体类上的@Entity 注解可以替换为@Table 注解。（　　）

5. Spring Boot 官方对 MyBatis 进行了整合,进一步简化了使用 MyBatis 进行数据的操作。（　　）

三、填空题

1. ＿＿＿＿注解表示该类是一个 MyBatis 接口文件,并保证能够被 Spring Boot 自动扫描到 Spring 容器中。

2. Spring Boot 整合 JPA 编写 Repository 接口文件数据修改方法时,除了@Query 注解外,还必须添加＿＿＿＿注解表示数据修改。

3. 在 mapper.xml 映射文件中,<mapper>标签的 namespace 属性值为对应接口文件的＿＿＿＿。

4. 使用 Spring Data JPA 自定义 Repository 接口类时,必须继承＿＿＿＿接口。

5. MyBatis 可以使用简单的 XML 或＿＿＿＿来配置和映射原生信息。

第5章

Spring Boot安全管理

本章学习目标
- 理解 Spring Security 的基本概念。
- 掌握安全管理效果测试方法。
- 熟悉自定义用户认证过程。
- 熟悉自定义用户授权管理过程。

根据党的二十大报告,网络安全作为网络强国、数字中国的底座,将在未来的发展中承负重担,是我国现代化产业体系中不可或缺的部分。

在实际开发中,一些应用通常要考虑到安全性问题。例如,对于一些重要的操作,有些请求需要用户验明身份后才可以执行,还有一些请求需要用户具有特定权限才可以执行。这样做不仅可以用来保护项目安全,而且可以控制项目访问效果。

本章将针对 Spring Boot 安全管理进行讲解,并进而体会网络安全的重要性。

5.1 Spring Security 简介

5.1.1 什么是 Spring Security

安全可以说是公司的红线,一般项目都有严格的认证和授权操作,在 Java 开发领域常见的安全框架有 Shiro 和 Spring Security。Shiro 是一个轻量级的安全管理框架,提供了认证、授权、会话管理、密码管理和缓存管理等功能。Spring Security 是一个相对复杂的安全管理框架,功能比 Shiro 更加强大,权限控制细粒度更高,对 OAuth2 的支持也更好。由于 Spring Security 源自 Spring 家族,因此可以和 Spring 框架无缝整合,特别是 Spring Boot 中提供的自动化配置方案,可以让 Spring Security 的使用更加便捷。

Spring Security 是 Spring 社区的一个顶级项目,也是 Spring Boot 官方推荐使用的安全框架。除了常规的认证(authentication)和授权(authorization)外,Spring Security 还提供了诸如 ACLs、LDAP、JAAS 和 CAS 等高级特性,以满足复杂场景下的安全需求。

Spring Security 应用级别的安全主要包含两个主要部分,即登录认证(login authentication)和访问授权(access authorization)。首先在用户登录时传入登录信息,登录验证器完成登录认证并将登录认证好的信息存储到请求上下文;然后进行其他操作,如接口访问、方法调用等,权限认证器从上下文中获取登录认证信息;最后根据认证信息获取权限信息,通过权限信息和特定的授权策略决定是否授权。

简单来说，Spring Security 核心功能只做以下几件事情：

(1) 在系统初始化时，告诉 Spring Security 访问路径所需要的对应权限。

(2) 在登录时，告诉 Spring Security 真实用户名和密码。

(3) 在登录成功时，告诉 Spring Security 当前用户具备的权限。

(4) 在用户访问接口时，Spring Security 已经知道用户具备的权限，也知道访问路径需要的对应权限，可以自动判断该用户能否访问。

5.1.2 为什么要使用 Spring Security

在项目开发中，安全框架多种多样，那么为什么要选择 Spring Security 作为微服务开发的安全框架呢？Java EE 有另一个优秀的安全框架 Apache Shiro，Apache Shiro 框架在企业级的项目开发中非常受欢迎，一般在单体项目中使用；在微服务架构中目前却是无能为力的。

选择 Spring Security 的原因之一是它来自于 Spring Resource 社区，采用了注解的方式来控制权限，熟悉 Spring 框架者很容易上手。另外一个原因是 Spring Security 很容易应用在 Spring Boot 工程中，也易于集成到 Spring Cloud 构建的微服务项目中。

Spring Security 是一个能够为基于 Spring 的企业应用系统提供声明式的安全访问控制解决方案的安全框架。它提供了一组可以在 Spring 应用上下文中配置的 Bean，充分利用了 Spring IoC(Inversion of Control，控制反转)、DI(Dependency Injection，依赖注入)和 AOP(Aspect Oriented Programming，面向切面编程)功能，为应用系统提供声明式的安全访问控制功能，减少了为企业系统安全控制编写大量重复代码的工作。

5.1.3 Spring Security 的核心类

1. SecurityContext

SecurityContext 中包含当前正在访问系统的用户的详细信息，它有以下两种方法。

- getAuthentication()：获取当前经过身份验证的主题或请求令牌。
- setAuthentication()：更改或删除当前已经验证的主体身份验证信息。

SecurityContext 的信息是由 SecurityContextHolder 来处理的。

2. SecurityContextHolder

SecurityContextHolder 是最基本的对象，保存当前会话用户认证、权限和鉴权等核心数据。SecurityContextHolder 默认使用 ThreadLocal 策略来存储认证信息、与线程绑定的策略等。在用户退出时，它会自动清除当前线程的认证信息。

最常用的方法是 getContext()方法，用来获得当前的 SecurityContext。该方法使用 Authentication 对象来描述当前用户的相关信息。SecurityContextHolder 持有的是当前用户的 SecurityContext，而 SecurityContext 持有的是代表当前用户相关信息的 Authentication 的引用。这个 Authentication 对象不需要自己创建，Spring Security 会自动创建，然后赋值给当前的 SecurityContext。

在程序的任何地方，可以通过如下方式获取到当前用户的用户名。

```
public String getCurrentUsername(){
    Object principal = SecurityContextHolder.getContext().getAuthentication().getPrincipal();
```

```
        if(principal instanceof UserDetails){
            return ((UserDetails) principal).getUsername();
        }
        if(principal instanceof Principal){
            return ((Principal) principal).getName();
        }
        return String.valueOf(principal);
}
```

getAuthentication()方法会返回认证信息。

getPrincipal()方法会返回身份信息,它是UserDetails对身份信息的封装。

获取当前用户的用户名,最简单的方式如下:

```
public String getCurrentUsername(){
    return SecurityContextHolder.getContext().getAuthentication().getName();
}
```

3. PrividerManager

PrividerManager会维护一个认证列表,用来处理不同认证方式的认证,这是因为系统可能会存在多种认证方式,如手机号、用户名密码、邮箱等。如果认证结果不是null,则说明成功,存在SecurityContext中;如果认证结果是null,则说明不成功,抛出ProviderNotFoundException异常。

4. DaoAuthenticationProvider

DaoAuthenticationProvider是AuthenticationProvider最常用的实现,用来获取用户提交的用户名和密码,并进行正确性比对。如果比对正确,则返回数据库中的用户信息。

5. UserDetails

UserDetails是Spring Security的用户实体类,包含用户名、密码、权限等信息。Spring Security默认实现了内置的User类,供Spring Security安全认证使用,也可以自己实现。

UserDetails接口和Authentication接口很类似,都拥有username和authorities。一定要区分清楚Authentication中的getCredentials()与UserDetails中的getPassword(),前者是用户提交的密码凭证,不一定正确或数据库不一定存在;后者是用户正确的密码。认证器要进行比对的就是两者是否相同。

UserDetails的实现代码如下:

```
public interface UserDetails extends Serializable {
    Collection<? extends GrantedAuthority> getAuthorities();
    String getPassword();
    String getUsername();
    boolean isAccountNonExpired();
    boolean isAccountNonLocked();
    boolean isCredentialsNonExpired();
    boolean isEnabled();
}
```

6. UserDetailsService

用户信息通过UserDetailsService接口加载。该接口的唯一方法是loadUserByUsername (String username),用来根据用户名加载相关信息。这个方法返回值是UserDetails接口,其中包含了用户的信息,包括用户名、密码、权限、是否启用等。

UserDetailsService 的实现代码如下：

```
public interface UserDetailsService {
    UserDetails loadUserByUsername(String var1) throws UsernameNotFoundException;
}
```

7. Authentication

Authentication 是建立系统使用者信息(principal)的过程。用户认证一般要求用户提供用户名和密码，系统通过用户名密码完成认证通过或拒绝。Spring Security 支持主流认证方式，包括 HTTP 基本认证、表单验证、摘要认证、OpenID 和 LDAP 等。除了利用提供的认证外，还可以编写自己的 Filter(过滤器)，以保证并非基于 Spring Security 的验证系统的操作。

用户认证的验证步骤如下：

(1) 用户使用用户名和密码登录。

(2) 过滤器获取到用户名和密码，然后封装成 Authentication。

(3) AuthenticationManager 认证 token(Authentication 的实现类传递)。

(4) AuthenticationManager 认证成功，返回一个封装了用户权限信息的 Authentication 对象，建立用户的上下文信息(如角色列表等)。

(5) Authentication 对象赋值给当前的 SecurityContext，建立这个用户的安全上下文(通过调用 SecurityContextHolder.getContext().setAuthentication()实现)。

(6) 用户进行一些受到访问控制机制保护的操作，访问控制机制会依据当前安全上下文信息检查这个操作所需要的权限。

Authentication 的实现代码如下：

```
public interface Authentication extends Principal, Serializable {
    //权限列表,通常是代表权限的字符串集合
    Collection<? extends GrantedAuthority> getAuthorities();
    //密码,认证之后会移出,用来保证安全性
    Object getCredentials();
    //请求的细节参数
    Object getDetails();
    //核心身份信息,一般返回 UserDetails 的实现类
    Object getPrincipal();
    boolean isAuthenticated();
    void setAuthenticated(boolean var1) throws IllegalArgumentException;
}
```

8. Authorization

在一个系统中，不同用户所具有的权限是不同的。系统会为不同的用户分配不同的角色，而每个角色则对应一系列的权限。

对 Web 资源的保护，最好的办法是使用过滤器。对方法调用的保护，最好的方法是使用 AOP。

5.2 安全管理效果测试

视频讲解

Spring Boot 针对 Spring Security 提供了自动化配置方案，因此可以使 Spring Security 非常容易地整合进 Spring Boot 项目中，这也是在 Spring Boot 项目中使用 Spring Security

的优势。

下面通过示例讲解 Spring Boot 实现安全管理的基本方法。

1. 添加依赖

以项目 springboot0406 为基础,创建项目 springboot0501,在 Maven 中添加依赖,以便启用安全管理。

```xml
<dependency>
    <groupId>org.springframework.boot</groupId>
    <artifactId>spring-boot-starter-security</artifactId>
</dependency>
```

2. 添加图书首页

添加网上图书的首页 index.html,该页面中通过标签分类展示了书店前台和书店后台两个选项,并且这两个选项都通过<a>标签连接到了具体的页面。

```html
<!DOCTYPE html>
<html xmlns="http://www.w3.org/1999/xhtml" xmlns:th="http://www.thymeleaf.org"
      xmlns:sec="http://www.thymeleaf.org/extras/spring-security">
<head>
    <meta http-equiv="Content-Type" content="text/html; charset=UTF-8">
    <title>网上书店</title>
</head>
<body>
    <h1 align="center">欢迎进入网上书店</h1>
    <div>
        <ul>
            <li><a th:href="@{/loginbooks}">书店前台</a></li>
            <li><a th:href="@{/loginadmin}">书店后台</a></li>
        </ul>
    </div>
</body>
</html>
```

3. 启动项目测试

接下来启动项目,启动成功后,执行"http://localhost:8080/"访问项目首页,会自动跳转到一个新的登录链接页面"http://localhost:8080/login",这说明在项目中添加 spring-boot-starter-security 依赖启动器后,项目实现了 Spring Security 的自动化配置,并且具有了一些默认的安全管理功能。另外,项目会自动跳转到登录页面,这个登录页面是由 Spring Security 提供的,如图 5-1 所示。

图 5-1 项目首页访问效果

当在 Spring Security 提供的默认登录页面"/login"中输入错误的登录信息后,会重定向到"/login?error"页面并显示出错误信息,如图 5-2 所示。

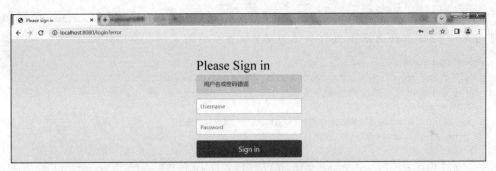

图 5-2 项目登录错误效果

需要说明的是，在 Spring Boot 项目中加入安全依赖启动器后，Security 会默认提供一个可登录的用户信息。该用户信息的用户名为 user，密码会随着项目的每次启动随机生成并打印在控制台上。查看项目启动日志，得到密码如图 5-3 所示。

```
Using generated security password: ee4a1590-8a18-4332-9008-81bc3586b6d4
```

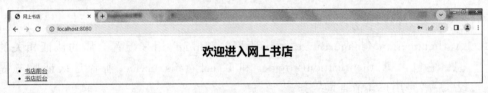

图 5-3 查看密码

在登录页面输入正确的用户名和密码，项目登录成功效果如图 5-4 所示。

图 5-4 项目登录成功效果

可以发现这种默认安全管理方式存在诸多问题，例如，只有唯一的默认登录用户 user、密码随机生成且过于暴露、登录页面及错误提示页面不是预想的等。

如果开发者对默认的用户名和密码不满意，可以在 application. properties 中配置默认的用户名、密码和用户角色，配置方式如下：

```
spring.security.user.name = zhangsan
spring.security.user.password = 123456
spring.security.user.roles = admin
```

再次启动项目，项目启动日志就不会打印出随机生成的密码了。用户可以直接使用配置好的用户名和密码登录，登录成功后，用户还会具有一个角色——admin。

5.3 自定义用户认证

通过自定义 WebSecurityConfigurerAdapter 类型的 Bean 组件，可以完全关闭 Security 提供的 Web 应用默认安全配置，但是不会关闭 UserDetailsService 用户信息自动配置类。如果要关闭 UserDetailsService 默认用户信息配置，可以自定义 UserDetailsService、AuthenticationProvider 或 AuthenticationManager 类型的 Bean 组件。另外，可以通过自定义 WebSecurityConfigurerAdapter 类型的 Bean 组件覆盖默认访问规则。Spring Boot 提供了很多方便的方法，可用于覆盖请求映射和静态资源的访问规则。

通过重写抽象接口 WebSecurityConfigurerAdapter，再加上注解@Configuration，就可以通过重写 configure 方法配置所需要的安全配置。自定义适配器的代码如下。

```
@Configuration
public class SecurityConfig extends WebSecurityConfigurerAdapter {
  //通常用于设置忽略权限的静态资源
  @Override
  public void configure(WebSecurity web) throws Exception {
    super.configure(web);
  }
  //通过 HTTP 对象的 authorizeRequests()方法定义 URL 访问权限，默认为 formLogin()提供一个简单的登录验证页面
  @Override
  protected void configure(HttpSecurity http) throws Exception {
    super.configure(http);
  }
  //通过 auth 对象的方法添加身份验证
  @Override
  protected void configure(AuthenticationManagerBuilder auth) throws Exception {
    super.configure(auth);
  }
}
```

自定义适配器的配置方法如下。

- AuthenticationManagerBuilder：认证相关 builder，用来配置全局的认证相关的信息。它包含 AuthenticationProvider 和 UserDetailsService，前者是认证服务提供者，后者是用户详情查询服务。
- HttpSecurity：进行权限控制规则相关配置。
- WebSecurity：进行全局请求忽略规则配置、HttpFirewall 配置、debug 配置和全局 SecurityFilterChain 配置。

视频讲解

5.3.1 内存身份认证

in-Memory Authentication（内存身份认证）是最简单的身份认证方式，主要用于 Security 安全认证体验和测试。在自定义内存身份认证时，只需要在重写的 configure (Authentication ManagerBuilder auth)方法中定义测试用户即可。下面通过 Spring Boot 整合 Spring Security 实现内存身份认证。

（1）创建内存身份认证。

打开项目 springboot0501，创建 SecurityConfig 类并继承 WebSecurityConfigurerAdapter，在重写的 configure(AuthenticationManagerBuilder auth)方法中使用内存身份认证。

```
@Configuration
public class SecurityConfig extends WebSecurityConfigurerAdapter {
    @Bean
    public PasswordEncoder passwordEncoder() {
        return new BcryptPasswordEncoder();
    }
    @Override
    protected void configure(AuthenticationManagerBuilder auth) throws Exception {
        User.UserBuilder builder = User.builder().passwordEncoder(passwordEncoder()::
```

```
encode);
        auth.inMemoryAuthentication().withUser(builder.username("zhangsan").password
                    ("123456").roles("common"));
        auth.inMemoryAuthentication().withUser(builder.username("lisi").password
                    ("654321").authorities("ROLE_vip", "ROLE_common"));
    }
}
```

自定义类继承 WebSecurityConfigurerAdapter 类，重写 configure 方法并在其中增加两个用户，配置用户名、密码和角色。需要注意的是，此处要额外设置密码的加密方式，否则在实际登录时会发现，即便输入了正确的用户名密码，也会提示登录失败。这是因为从 Spring Security 5 开始，自定义用户认证必须设置密码编码器用于保护密码，否则控制台会出现异常错误。Spring Security 提供了多种密码编码器，包括 BcryptPasswordEncoder、Pbkdf2PasswordEncoder 和 ScryptPasswordEncoder 等，密码设置不限于本例中的 BcryptPasswordEncoder 密码编码器。

（2）运行程序，效果如图 5-1 所示。

5.3.2　JDBC 身份认证

视频讲解

JDBC Authentication（JDBC 身份认证）通过 JDBC 连接数据库对已有用户身份进行认证。下面通过 Spring Boot 整合 Spring Security 实现 JDBC 身份认证。

（1）数据准备。

JDBC 身份认证的本质是使用数据库中已有的用户信息在项目中实现用户认证服务，所以需要提前准备好相关数据。这里使用之前创建的名为 book 的数据库，在该数据库中修改之前创建的 3 个表 user、role 和 user_role，并插入几条测试数据，分别如图 5-5～图 5-7 所示。

图 5-5　用户表

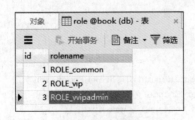

图 5-6　权限表

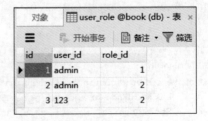

图 5-7　用户权限表

在使用 JDBC 身份认证创建用户/权限表时，需要注意以下几点。

① 在创建用户表 user 时，用户名 id 必须唯一，因为 Security 在进行用户查询时是通过 id 定位是否存在唯一用户的。

② 在创建用户表 user 时，必须额外定义一个 tinyinit 类型的字段（对应 boolean 类型的

属性,如示例中的 valid),用于校验用户身份是否合法(默认都是合法的)。

③ 在初始化用户表 user 数据时,插入的用户密码 password 必须是对应编码器编码后的密码,如示例中的密码就是加密后的形式(对应的原始密码为 123456)。因此,在自定义配置类中进行用户密码查询时,必须使用与数据库密码统一的密码编码器进行编码。

④ 在初始化角色表 role 数据时,角色 rolename 值必须带有"ROLE_"前缀,而默认的用户角色值则是对应权限值去掉"ROLE_"前缀。这是因为之前的 role 都是通过 springsecurity 的 api 进行赋值的,它会自行加上该前缀。

(2) 添加 JDBC 连接数据库的依赖驱动启动器。

打开项目 springboot0501 中的 pom.xml 文件,在该文件中添加对应的依赖。

```xml
<dependency>
    <groupId>org.springframework.boot</groupId>
    <artifactId>spring-boot-starter-data-jpa</artifactId>
</dependency>
<dependency>
    <groupId>mysql</groupId>
    <artifactId>mysql-connector-java</artifactId>
    <version>5.1.30</version>
</dependency>
```

(3) 进行数据库连接配置。

在项目的全局配置文件 application.properties 中编写对应的数据库连接配置。

```
spring.datasource.url=jdbc:mysql://localhost:3306/book
spring.datasource.username=root
spring.datasource.password=123456
spring.datasource.driver-class-name=com.mysql.jdbc.Driver
```

(4) 使用 JDBC 进行身份认证。

打开项目 springboot0501,创建 SecurityConfig 类并继承 WebSecurityConfigurerAdapter,在重写的 configure(AuthenticationManagerBuilder auth) 方法中使用 JDBC 身份认证。

```java
@Configuration
public class SecurityConfig extends WebSecurityConfigurerAdapter {
    @Bean
    public PasswordEncoder passwordEncoder() {
        return new BCryptPasswordEncoder();
    }
    @Autowired
    private DataSource datasource;
    @Override
    protected void configure(AuthenticationManagerBuilder auth) throws Exception {
        PasswordEncoder encoder = new BCryptPasswordEncoder();
        String userSql = "select id,password,valid from user where id = ?";
        String authoritySql = "SELECT u.id,r.rolename FROM user AS u ,role AS r ,user_role AS ur WHERE u.id = ? AND u.id = ur.user_id AND r.id = ur.role_id";
        auth.jdbcAuthentication().passwordEncoder(encoder)
                .dataSource(datasource)
                .usersByUsernameQuery(userSql)
                .authoritiesByUsernameQuery(authoritySql);
    }
}
```

需要注意的是,在定义用户查询的 SQL 语句时,必须返回用户名 id、密码 password、是否为有效用户 valid 这 3 个字段信息;在定义权限查询的 SQL 语句时,必须返回用户名 id、角色 rolename 这两个字段信息。否则,登录时输入正确的用户信息会出现 PreparedStatementCallback 的 SQL 异常错误信息。

实际项目中并不会把密码明文存储在数据库中。默认使用的 PasswordEncoder 要求数据库中的密码格式为{id}password,它会根据 id 去判断密码的加密方式,但是一般不会采用这种方式。通常需要替换 PasswordEncoder,使用 Spring Security 提供的 BCryptPasswordEncoder,将 BCryptPasswordEncoder 对象注入 Spring 容器中,Spring Security 就会使用该 PasswordEncoder 进行密码校验。

(5)运行程序,效果如图 5-1 所示。

5.3.3　UserDetailsService 身份认证

视频讲解

采用配置文件的方式可以从数据库中读取用户进行登录。虽然该方式相较于静态账号密码的方式更加灵活,但是将数据库的结构暴露在明显的位置上,绝对不是一个明智的做法。接下来将通过 UserDetailsService 接口实现身份认证。

Spring Security 中进行身份验证的是 AuthenticationManager 接口,ProviderManager 是该接口的一个默认实现,但它并不用来处理身份认证,而是委托给配置好的 AuthenticationProvider,每个 AuthenticationProvider 会轮流检查身份认证,在检查后返回 Authentication 对象或抛出异常。

验证身份就是加载响应的 UserDetails,比对是否与用户输入的账号、密码、权限等信息匹配。该步骤由实现 AuthenticationProvider 的 DaoAuthenticationProvider(它利用 UserDetailsService 验证用户名、密码和授权)处理,包含 GrantedAuthority 的 UserDetails 对象在构建 Authentication 对象时填入数据,如图 5-8 所示。

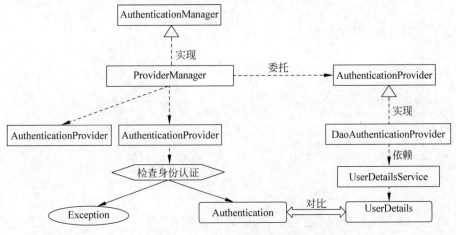

图 5-8　UserDetailsService 身份认证

正常的登录流程可能就是前端传过来账号密码,后端收到账号密码,直接通过账号密码的两个条件进行数据库查询,查询到就登录成功,查询不到就登录失败。SpringSecurity 并非如此,其流程如下:

(1)前端将账号和密码传送给后端(这里直接以明文举例)。

(2) 后端通过输入账号获取用户信息(获取不到则证明该账号不存在)。

(3) 获取信息后进行密码比对,看看是否正确(该过程称为身份认证)。

UserDetailsService 接口的主要作用就是该流程的第二个步骤。

下面通过 Spring Boot 整合 Spring Security 实现 UserDetailsService 身份认证。

(1) 创建数据库实体类。

User 类:

```
@Entity(name = "user")
@Data
public class User implements Serializable {
  @Id
  private String id;
  private String password, username, address, email, phone;
  @ManyToMany
  @JoinTable(name = "user_role", joinColumns = @JoinColumn(name = "user_id"),
           inverseJoinColumns = @JoinColumn(name = "role_id"))
  private List<Role> roles;
}
```

Role 类:

```
@Data
@Entity(name = "role")
public class Role implements Serializable {
  @Id
  @GeneratedValue(strategy = GenerationType.IDENTITY)
  private Integer id;
  private String rolename;
}
```

(2) 创建 UserRepository。

```
public interface UserRepository extends JpaRepository<User, String> {
}
```

(3) 定义查询用户及角色信息的服务接口。

```
@Service
public class UserServiceImpl implements UserService {
  @Autowired
  private UserRepository dao;
  @Cacheable(cacheNames = "user", key = "#id")
  @Override
  public User findById(String id) {
      User user = dao.getById(id);
      return user;
  }
}
```

(4) 定义 UserDetailsService 接口,用于封装认证用户信息。

UserDetailsService 接口只有一个方法,该方法返回值为 UserDetails,UserDetails 类是系统默认的用户"主体"。在用户登录时会访问该方法,并将登录传入的用户名以参数形式进行传送。为此需要做的就是实现 UserDetailsService 接口,然后重写这个方法,并返回一个 UserDetails 对象。

定义 UserDetailsService 接口的步骤如下。

① 根据登录用户名查询数据库，判断该用户是否存在。
② 将从数据库查询出的账号密码封装到 UserDetails 对象中，作为方法返回值返回。

```
@Service
@Qualifier
public class UserDetailsServiceImpl implements UserDetailsService {
    @Autowired
    private UserService service;
    @Override
    public UserDetails loadUserByUsername(String username) throws UsernameNotFoundException {
      it.com.boot.springboot05_01.po.User user = service.findByid(username);
      if (user!= null) {
      List < SimpleGrantedAuthority > authorities = new ArrayList <>();
      for (Role role : user.getRoles()) {
        SimpleGrantedAuthority simpleGrantedAuthority = new SimpleGrantedAuthority(role.getRolename());
        authorities.add(simpleGrantedAuthority);
      }
      UserDetails userDetails = new User(username, user.getPassword(), authorities);
      return userDetails;
    }
      else
      throw new UsernameNotFoundException("当前用户不存在!");
    }
}
```

需要注意的是，在 UserDetailsServiceImpl 业务处理类获取 User 实体类时，必须对当前用户进行非空判断。这里使用 throw 进行异常处理，如果查询的用户为空，throw 会抛出 UsernameNotFoundException 的异常。如果没有使用 throw 异常处理，Security 将无法识别，导致程序整体报错。

UserDetails（位于 org.springframework.security.core.userdetails 包下）主要与用户信息有关，该接口是提供用户信息的核心接口。该接口仅能实现存储用户的信息，后续会将该接口提供的用户信息封装到认证对象 Authentication 中。

UserDetails 的子类 User（位于 org.springframework.security.core.userdetails 包下）是用户类，用于存放待认证的用户信息。如从数据库查出的用户信息和从缓存中获取的用户信息都可以封装到此类中，最终交给 SpringSecurity 进行用户信息认证。UserDetails 实体类的结构如图 5-9 所示。

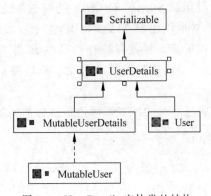

图 5-9　UserDetails 实体类的结构

(5) 使用 UserDetailsService 进行身份认证。

打开项目 springboot0501, 创建 SecurityConfig 类并继承 WebSecurityConfigurerAdapter, 在重写的 configure(AuthenticationManagerBuilder auth) 方法中使用 JDBC 身份认证。

```
@Configuration
public class SecurityConfig extends WebSecurityConfigurerAdapter {
    @Bean
    public PasswordEncoder passwordEncoder() {
        return new BCryptPasswordEncoder();
    }
    @Autowired
    @Qualifier
    private UserDetailsService userDetailsService;
    @Override
    protected void configure(AuthenticationManagerBuilder auth) throws Exception {
        PasswordEncoder encoder = new BCryptPasswordEncoder();
        auth.userDetailsService(userDetailsService).passwordEncoder(encoder);
    }
}
```

(6) 运行程序, 效果如图 5-1 所示。

5.4 自定义用户授权管理

5.3 节的案例只是做了请求认证, 并没有对权限进行控制。下面基于 5.3 节的案例进行权限控制处理。

所谓权限控制, 以一个学校图书馆的管理系统为例进行分析。如果是普通学生登录, 能看到借书还书相关的功能, 但不能看到或使用添加书籍信息、删除书籍信息等功能。如果是图书馆管理员登录, 就能看到并使用添加书籍信息、删除书籍信息等功能。

总结起来就是不同的用户可以使用不同的功能, 这就是权限系统要实现的效果。

在权限控制的实现过程中, 不能只依赖前端判断用户的权限。因为如果只是这样, 当有人知道了对应功能的接口地址时, 就可以不通过前端而直接发送请求实现相关功能操作。所以还需要在后台进行用户权限的判断, 判断当前用户是否有相应的权限, 必须具有所需权限才能进行相应的操作。

权限控制系统一般都会分为角色控制和菜单控制, 菜单控制又分为页面菜单和按钮控制。所谓按钮控制, 就是指定 Java 某个接口必须具备什么角色, 或者具备按钮权限才可以访问。

在一些比较早的项目中没用到 Spring Security, 当时的做法如下:

前端在登录时访问后端, 获取该用户有哪些权限, 这个权限包含菜单权限和按钮权限, 当不具备某个按钮权限时直接屏蔽。拦截器只有一层拦截, 即登录成功的人就能访问所有接口, 对于普通人来说该功能只是程序员设置的假象。一旦登录成功, 只要知道接口名称, 便可以通过接口直接访问。

Spring Security 只需要通过简单的配置即可避免这样的问题。

5.4.1 授权基本流程

在 Spring Security 中, 会使用默认的 FilterSecurityInterceptor 进行权限校验。FilterSecurityInterceptor 会获取 SecurityContextHolder 中的 Authentication, 然后获取其

中的权限信息,包括当前用户是否拥有访问当前资源所需的权限。所以在项目中只需要把当前登录用户的权限信息存入 Authentication,然后设置资源所需要的权限即可。

在实际生产中,网站访问多是基于 HTTP 请求的,通过重写 WebSecurityConfigurerAdapter 类的 config(HttpSecurity http)方法可以对基于 HTTP 的请求访问进行控制。configure (HttpSecurity http)方法的参数类型是 HttpSecurity 类,HttpSecurity 类提供了 HTTP 请求的限制权限、Session 管理配置、CSRF 跨站请求问题等方法。

下面通过对 configure(HttpSecurity http)方法的剖析,分析自定义用户访问控制的实现过程。

1. 访问控制 URL 匹配

配置类中的 httpSecurity.authorizeRequests()主要是对 URL 进行控制,也就是授权(访问控制)。httpSecurity.authorizeRequests()支持连缀写法,可以有很多 URL 匹配规则和权限控制方法,这些内容进行各种组合就形成了 Spring Security 中的授权。

在所有匹配规则中取所有规则的交集。配置顺序会影响之后的授权效果,越是具体的配置越应该放在前面,越是模糊的配置越应该放到后面。

(1) antMatcher()。

antMatcher()的匹配规则如下:

① ? 匹配一个字符

② * 匹配 0 个或多个字符

③ ** 匹配 0 个或多个目录

在实际项目中经常需要放行所有静态资源,下面的代码表示放行 js 文件夹下的所有脚本文件。

.antMatchers("/js/**").permitAll()

还有一种配置方式是放行所有.js 文件。

antMatchers("/**/*.js").permitAll()

(2) anyRequest()。

在之前的认证过程中使用的是 anyRequest(),表示匹配所有的请求。一般情况下都会使用该方法,并设置全部内容都需要进行认证。

代码示例:

anyRequest().authenticated();

2. 内置访问控制

Spring Security 在匹配 URL 后会调用 permitAll(),表示不需要认证即可随意访问。Spring Security 提供了多种内置控制,其底层都是基于 access 进行实现的。

(1) permitAll()。

permitAll()表示所匹配的 URL 允许任何人访问。

```
public ExpressionUrlAuthorizationConfigurer<H>.ExpressionInterceptUrlRegistry permitAll() {
    return this.access("permitAll");
}
```

(2) authenticated()。

authenticated()表示所匹配的 URL 需要被认证才能访问。

```java
public ExpressionUrlAuthorizationConfigurer<H>.ExpressionInterceptUrlRegistry
authenticated() {
    return this.access("authenticated");
}
```

(3) anonymous()。

anonymous()表示可以匿名访问匹配的 URL。与 permitAll()效果类似，但设置为 anonymous()的 URL 会执行 filter 链中的语句。

```java
public ExpressionUrlAuthorizationConfigurer<H>.ExpressionInterceptUrlRegistry
anonymous() {
    return this.access("anonymous");
}
```

(4) denyAll()。

denyAll()表示所匹配的 URL 不允许被访问。

```java
public ExpressionUrlAuthorizationConfigurer<H>.ExpressionInterceptUrlRegistry denyAll() {
    return this.access("denyAll");
}
```

(5) rememberMe()。

rememberMe()表示被"记住"的用户允许访问。

```java
public ExpressionUrlAuthorizationConfigurer<H>.ExpressionInterceptUrlRegistry
rememberMe() {
    return this.access("rememberMe");
}
```

(6) fullyAuthenticated()。

fullyAuthenticated()表示不被"记住"的用户才可以访问。

```java
public ExpressionUrlAuthorizationConfigurer<H>.ExpressionInterceptUrlRegistry
fullyAuthenticated() {
    return this.access("fullyAuthenticated");
}
```

3. 角色权限判断

除了之前讲解的内置权限控制外，Spring Security 还支持很多其他权限控制。这些方法一般都用于用户已经被认证后，判断用户是否具有特定的要求。

底层也是调用 access(参数)，参数正好是调用的方法名。需要注意的是，判断角色会给调用方法参数前面添加 ROLE_，这也是为什么正常调用方法时角色不允许以 ROLE_开头命名的原因。

(1) hasAuthority(String)。

判断用户是否具有特定的权限，用户的权限是在自定义登录逻辑中创建 User 对象时指定的。

```java
List<String> listPermission = userMapper.selectPermissionByUsername(username);
List<SimpleGrantedAuthority> listAuthority = new ArrayList<SimpleGrantedAuthority>();
for(String permisssion : listPermission){
    listAuthority.add(new SimpleGrantedAuthority(permisssion));
}
```

在配置类中通过 hasAuthority("admin")设置具有 admin 权限时才能访问,代码如下。

```
.antMatchers("/main1.html").hasAuthority("admin")
```

(2) hasAnyAuthority(String)。

如果用户具备给定权限中的某一个就允许访问。

下面代码中由于大小写和用户的权限不相同,所以用户无权访问/main1.html。

```
.antMatchers("/main1.html").hasAnyAuthority("adMin","admiN")
```

(3) hasRole(String)。

如果用户具备给定角色就允许访问,否则出现 403 错误提示。

参数取值来源于自定义登录逻辑 UserDetailsService 实现类中创建 User 对象时给 User 赋予的授权。在给用户赋予角色时,角色需要以 ROLE_开头,其后添加角色名称。例如,ROLE_abc,其中 abc 是角色名,ROLE_是固定的字符开头。

```
List<String> listPermission = userMapper.selectPermissionByUsername(username);
List<String> listRoles = userMapper.selectRoleByUsername(username);
for(String role: listRoles){
    listAuthority.add(new SimpleGrantedAuthority("ROLE_" + role));
}
```

在配置类中添加代码如下。

```
.antMatchers("/bjsxt").hasRole("abc")
```

(4) hasAnyRole(String …)。

如果用户具备给定角色的任意一个就允许访问。

5.4.2 自定义登录页面

虽然 Spring Security 提供了登录页面,但是用户在实际项目中大多喜欢使用自己的登录页面。因此 Spring Security 不仅提供了登录页面,而且支持用户自定义登录页面。实现过程也比较简单,只需要修改配置类即可,具体步骤如下。

视频讲解

(1) 编写登录页面。

编写登录页面,其中<form>的 action 可以不编写对应控制器。

```
<form class="form-signin" th:action="@{/tologin}" th:method="post">
    <img class="mb-4" th:src="@{/login/img/logo.gif}" width="72px" height="72px">
    <h1 class="h3 mb-3 font-weight-normal">请登录</h1>
    <!-- 用户登录错误信息提示框 -->
    <div th:if="${param.error}" style="color: red;height: 40px;text-align: left;font-size: 1.1em">
        <img th:src="@{/login/img/loginError.jpg}" width="20px">用户名或密码错误,请重新登录!
    </div>
    <input type="text" name="name" class="form-control" placeholder="用户名" required="" autofocus="">
    <input type="password" name="pwd" class="form-control" placeholder="密码" required="">
    <button class="btn btn-lg btn-primary btn-block" type="submit">登录</button>
    <p class="mt-5 mb-3 text-muted">Copyright © 2019 - 2020 </p>
</form>
```

(2) 编写控制器。

在创建的 LoginController 类中添加一个跳转到登录页面 login.html 的方法。

```
@GetMapping("/toindex")
public String toIndex() {
    return "login";
}
```

在上述添加的 toIndex()方法中,配置了请求路径为/toindex 的 GET 请求,并向静态资源根目录下的 login.html 页面跳转。

Spring Security 默认向登录页面跳转时,采用的请求方式是 GET,请求路径是/login;如果要处理登录后的数据,默认采用的请求方式是 POST,请求路径是 login。

表单处理成功会跳转到一个地址,失败也会跳转到一个地址。在控制器类中添加控制器方法,方法映射路径为/fail。此处需要注意的是,如果是 POST 请求访问/fail,一旦返回值直接转发到 fail.html 中,即使有效果,控制台也会报警提示 fail.html 不支持 POST 访问方式。

```
@GetMapping("/fail")
public String fail(){
    return "fail";
}
```

(3) 创建登录失败页面 fail.html。

```
<!DOCTYPE html>
<html xmlns="http://www.w3.org/1999/xhtml">
<head>
    <meta http-equiv="Content-Type" content="text/html; charset=UTF-8">
    <title>网上书店</title>
</head>
<body>
<h1 align="center">欢迎进入网上书店</h1>
<div>
    操作失败,请重新登录.<a href="/toindex">跳转</a>
</div>
</body>
</html>
```

(4) 修改配置类。

修改配置类主要是设置哪个页面是登录页面,配置类需要继承 WebSecurityConfigurerAdapter,并重写 configure 方法。

```
@Override
protected void configure(HttpSecurity http) throws Exception {
    http.authorizeRequests().antMatchers("/").permitAll()
            .antMatchers("/login/**").permitAll()
            .anyRequest().authenticated();
    http.formLogin()
            //定义登录时用户名的 key,默认为 username
            .usernameParameter("name")
            //定义登录时用户密码的 key,默认为 password
            .passwordParameter("pwd")
            //登录页面表单提交地址,此地址可以不真实存在
```

```
                .loginProcessingUrl("/tologin")
        //登录成功之后跳转到哪个 URL
                .defaultSuccessUrl("/")
        //登录失败之后跳转到哪个 URL
                .failureUrl("/fail").permitAll()
        //登录页面
                .loginPage("/toindex").permitAll();
        /*
        http.formLogin()
                .usernameParameter("name")
                .passwordParameter("pwd")
                .loginProcessingUrl("/tologin")
                .defaultSuccessUrl("/")
                .failureUrl("/toindex?error")
                .loginPage("/toindex").permitAll();
        */
    }
```

failureUrl("/toindex?error")方法用来控制用户登录认证失败后的跳转路径,该方法默认参数为"/login?error"。其中,参数中的"/toindex"为向登录页面跳转的映射;error是一个错误标识,作用是登录失败后在登录页面进行接收判断,如 login.html 实例中的 ${param.error},这两者必须保持一致。

(5) 运行项目。

自定义登录页面如图 5-10 所示。输入正常的用户名和密码,单击"登录"按钮,跳转到首页,如图 5-11 所示。

图 5-10　自定义登录页面

图 5-11　网上书店首页

当输入用户名或密码有误时,如果配置文件中配置的是跳转到 fail.html,单击"登录"按钮,如图 5-12 所示。

图 5-12 登录失败页面

如果配置文件中配置的是跳转到登录页面,且同时携带一个错误标识 error,单击"登录"按钮,如图 5-13 所示。

图 5-13 错误提示

在传统项目中进行用户登录处理时,通常会查询用户是否存在,如果存在则登录成功,同时将当前用户放在 Session 中。

Spring Security 针对拦截的登录用户专门提供了一个 SecurityContextHolder 类,该类存储了当前与系统交互的用户信息。Spring Security 使用一个 Authentication 对象来表示这些信息。一般不需要自己创建这个对象,但是查找这个对象的操作对用户来说却十分常见。

```
SecurityContext context = SecurityContextHolder.getContext();
Authentication authentication = context.getAuthentication();
```

(6)修改配置文件,完成登录成功后,在后台获取用户名和权限信息。

```
@Override
protected void configure(HttpSecurity http) throws Exception {
    http.authorizeRequests().antMatchers("/").permitAll()
            .antMatchers("/login/**").permitAll()
            .anyRequest().authenticated();
    http.formLogin()
            .usernameParameter("name")
            .passwordParameter("pwd")
            .loginProcessingUrl("/tologin")
            .successHandler(new AuthenticationSuccessHandler() {
                @Override
                public void onAuthenticationSuccess(HttpServletRequest request, HttpServletResponse response, Authentication authentication) throws IOException, ServletException {
                    //第一种方法
```

```
                    System.out.println(authentication.getName());
                    authentication.getAuthorities().forEach(System.out::println);
                    System.out.println("------------------------");
                    //第二种方法
                    UserDetails userDetails = (UserDetails) authentication.getPrincipal();
                    System.out.println(userDetails.getUsername());
                    userDetails.getAuthorities().forEach(System.out::println);
                    response.sendRedirect("/");
                }
            })
            .failureUrl("/toindex?error")
            .loginPage("/toindex").permitAll();
}
```

（7）再次运行项目。

再次运行项目并正确登录后，在控制台输出当前登录的用户信息，如图 5-14 所示。

图 5-14　控制台输出登录用户信息

5.4.3　权限控制和注销

在 Spring Security 框架下实现用户的"退出"logout 的功能其实非常简单，具体实现步骤如下。

（1）修改配置类。

```
@Override
protected void configure(HttpSecurity http) throws Exception {
    http.logout();
}
```

（2）在 indexhtml 页面中增加一个注销按钮。

```
< form th:action = "@{/logout}" method = "post">
    < input type = "submit" value = "注销">
</form>
```

HttpSecurity 类的 logout()方法用来处理用户退出，它默认处理路径为/logout 的 POST 类型请求，同时也会清除 Session 和 Remember Me 等任何默认用户配置。

（3）运行测试，在登录成功后单击"注销"按钮，发现注销完成后会跳转到登录页面。

（4）进行个性化配置。

Spring Security 默认使用/logout 作为退出处理请求路径，将登录页面作为退出后的跳转页面。这符合绝大多数的应用的开发逻辑，但有的时候我们需要一些个性化设置。

```
@Override
protected void configure(HttpSecurity http) throws Exception {
    http.logout().logoutUrl("/mylogout").logoutSuccessUrl("/");
}
```

通过指定 logoutUrl 配置改变退出请求的默认路径,当然 HTML 退出按钮的请求 URL 也要修改,同时通过指定 logoutSuccessUrl 配置显式指定退出后的跳转页面。

(5) 运行项目。

在项目首页上方已经出现了新添加的用户退出链接,如图 5-15 所示。单击"注销"按钮后,发现跳转到首页。

图 5-15 用户注销

现在提出一个需求:对应不同权限的用户应该能够访问不同的页面请求。例如,sili 这个用户拥有 common 权限,所以只能查看书店前台而不能访问书店后台。

(6) 修改配置类,对书店前后台访问的控制器设置不同的权限。

```
@Override
protected void configure(HttpSecurity http) throws Exception {
    http.authorizeRequests().antMatchers("/").permitAll()
            .antMatchers("/login/**").permitAll()
            .antMatchers("/loginadmin").hasAnyRole("vip","vvip")
            .antMatchers("/loginbooks").hasRole("common")
            .anyRequest().authenticated();
}
```

(7) 运行项目。

以普通身份 zhaosi 访问书店前台,如图 5-16 所示。然后以该身份访问书店后台,此时由于身份权限的设置,页面会发出 403 拒绝响应,如图 5-17 所示。

图 5-16 以普通身份访问书店前台

接下来再提出几个需求:在用户没有登录时,导航栏上只显示登录按钮;在用户登录后,导航栏可以显示登录的用户信息及注销按钮;当用户只有普通权限时,登录后只能显示出书店前台的功能,而不显示书店后台的功能。这就是真实的网站情况。该如何完成需求

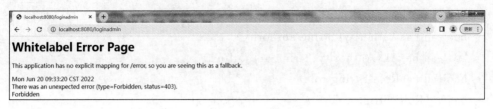

图 5-17　403 拒绝响应

呢？这里需要结合 Thymeleaf 中的一些功能。

（8）导入 Thymeleaf 和 Security 结合的 Maven 依赖。

<dependency>
　　<groupId>org.thymeleaf.extras</groupId>
　　<artifactId>thymeleaf-extras-springsecurity5</artifactId>
</dependency>

（9）修改前端页面 index.html。

① 导入命名空间。

<html xmlns="http://www.w3.org/1999/xhtml" xmlns:th="http://www.thymeleaf.org"
　　xmlns:sec="http://www.thymeleaf.org/extras/spring-security">

② 修改导航栏，增加认证判断。

<body>
<h1 align="center">欢迎进入网上书店</h1>
<div sec:authorize="isAnonymous()">
　　<h2 align="center">访问您好,请先<a th:href="@{/toindex}">登录</a></h2>
</div>
<div sec:authorize="isAuthenticated()">
　　<h2 align="center"><span sec:authentication="name"></span>你好,你的权限是<span sec:authentication="authorities"/>,你能够访问以下内容。
　　</h2>
　　<h2 align="center"><span sec:authentication="principal.username"></span>你好,你的权限是<span sec:authentication="principal.authorities"/>,你能够访问以下内容。</h2>
</div>
<form th:action="@{/mylogout}" method="post">
　　<input type="submit" value="注销">
</form>
<hr>
<div>
　　<ul>
　　　　<li sec:authorize="hasRole('common')"><a th:href="@{/loginbooks}">书店前台</a></li>
　　　　<li sec:authorize="hasAnyRole('vip','vvip')"><a th:href="@{/loginadmin}">书店后台</a></li>
　　</ul>
</div>
</body>

页面顶部通过"xmlns:sec"引入了 Security 安全标签库。常用的标签库如下。

- sec:authorize 权限
- sec:authentication 认证
- sec:authorize-url 不能直接使用，需要额外的配置（不建议使用该标签，因为该标签不支持 RESTful 风格）

- SecurityExpressionRoot 常用的表达式
- boolean hasAuthority(String authority)
- boolean hasAnyAuthority(String authorities)
- boolean hasRole(String role)
- hasAnyRole(String roles)
- Authentication getAuthentication()
- Object getPrincipal()
- boolean isAnonymous()
- boolean isAuthenticated()
- boolean isRememberMe()

下面对<div>模块的作用及内部属性进行详细说明。

sec:authorize="isAnonymous()"属性,判断用户是否未登录,只有匿名用户(未登录用户)才会显示登录链接提示。

sec:authorize="isAuthenticated()"属性,判断用户是否已登录,只有认证用户才会显示登录用户信息和注销链接等提示。

sec:authorize="hasRole('common')"属性,定义了只有角色为 common(对应权限 Authority 为 ROLE_common)且登录的用户才会显示书店前台列表信息。

sec:authorize="hasAuthority('ROLE_vip')"属性,定义了只有权限为 ROLE_vip(对应角色 Role 为 vip)且登录的用户才会显示书店后台列表信息。

sec:authorize="hasAnyRole('vip','vvip')"属性,定义了只有权限为 vip 或 vvip,且登录的用户才会显示书店后台列表信息。

Spring Security 首先利用 BeanWrapperImpl 封装了 Authentication 对象,然后调用 BeanWrapperImpl 的 getPropertyValue()方法获取 property 属性的值。而 BeanWrapperImpl 类能够通过 name-value 值对的方式对目标对象(Authentication 对象)进行属性(属性可以为嵌套属性)操作,所以 property 的取值可以是 Authentication 的直接属性或嵌套属性。

sec:authentication="name"和 sec:authentication="principal.username"这两个属性都是获取当前登录用户的用户名。

sec:authentication="authorities"和 sec:authentication="principal.authorities"这两个属性都是获取当前登录用户的所有角色。

(10) 运行项目。

当用户以匿名身份进行首页登录时,效果如图 5-18 所示。从图 5-18 可以看出,此次访问项目首页时,页面上方出现"请先登录"的链接,页面中不再显示书店前后台的链接,说明页面安全访问控制实现了效果。接着单击"登录"链接,输入正确的用户名和密码,登录成功

图 5-18 未登录访问首页效果

后跳转到项目首页。由于用户 zhaosi 只拥有 common 权限,所以只能显示书店前台选项,如图 5-19 所示。

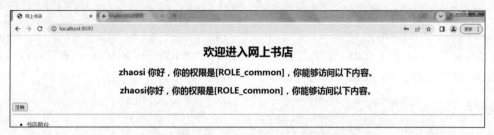

图 5-19　登录后访问首页效果

5.4.4　"记住我"及首页定制

视频讲解

现在网上书店的情况是,只要登录后关闭浏览器,再登录时就会要求重新登录。很多网站的有一个记住密码的功能,这个该如何实现呢?记住密码功能"记住我"的实现过程如图 5-20 所示。

图 5-20　"记住我"的实现过程

(1) 浏览器在发送用户请求时,会到 UsernamePasswordAuthenticationFilter 进行认证,该过滤器认证成功后会调用 RememberMeService 服务。

(2) RememberMeService 会生成一个 Token,并把这个 Token 写入浏览器的 Cookie 里面。注意此时在客户端的 Cookie 中,仅保存一个无意义的加密串(与用户名、密码等敏感数据无关)。同时 RememberMeService 会用 TokenRepository 将 Token 存入数据库中,因为是认证成功后进行存库操作,所以 SpringSecurity 会将用户名也写入库中。用户名与 Token 是一一对应的。

(3) 第二天再访问系统请求时会经过一个 RememberMeAuthenticationFilter 过滤器,这个过滤器的作用就是读取 Cookie 中的 Token 并将其交给 RememberMeService。RememberMeService 会根据 Token 查询数据库中是否有记录,有记录则将用户名取出并调用 UserDetailsServicer 将其放入 SecurityContext 中,此时登录成功。

下面继续完善前面的项目,并添加"记住我"这个功能。

(1) 创建一张记录 RememberMe 的表,用于记录令牌信息。

```
create table persistent_logins (username varchar(64) not null,
                series varchar(64) primary key,
                token varchar(64) not null,
                last_used timestamp not null);
```

这张表的名称和字段都是官方提供的固定模板，可以使用默认的 JDBC，即 JdbcTokenRepositoryImpl 来实现"记住我"功能。这张表也可以完全自定义，还可以使用系统提供的 JDBC 来操作。

（2）编写登录页面。

在 login.html 登录页面中添加"记住我"选项，其中"记住我"的选择框是 checkbox 类型的多选框，它的 name 属性可以选择默认值 rememberme。

```html
<div class="checkbox mb-3">
  <label>
    <input type="checkbox" name="rememberme">"记住我"选项
  </label>
</div>
```

（3）添加依赖。

Spring Security 在实现 RememberMe 功能时，其底层实现依赖 SpringJDBC，所以需要导入 SpringJDBC。由于此项目使用的是 JPA，所以此处导入 JPA 启动器，同时还需要添加 MySQL 驱动。

```xml
<dependency>
    <groupId>org.springframework.boot</groupId>
    <artifactId>spring-boot-starter-data-jpa</artifactId>
</dependency>
<dependency>
    <groupId>mysql</groupId>
    <artifactId>mysql-connector-java</artifactId>
    <version>5.1.30</version>
</dependency>
```

（4）配置数据源。

在 application.properties 中配置数据源，这在前面已经配置完成，这里就不再赘述了。

（5）编写配置文件。

配置 TokenRepository 并在 configure 中指定 RememberMe 需要的配置包含 TokenRepository 对象和 Token 过期时间。

```java
@Configuration
public class SecurityConfig extends WebSecurityConfigurerAdapter {
    @Autowired
    private DataSource datasource;
    @Override
    protected void configure(HttpSecurity http) throws Exception {
        http.rememberMe().rememberMeParameter("rememberme")
.tokenValiditySeconds(60).tokenRepository(persistentTokenRepository());
    }
    @Bean
    public PersistentTokenRepository persistentTokenRepository() {
        JdbcTokenRepositoryImpl tokenRepository = new JdbcTokenRepositoryImpl();
        tokenRepository.setDataSource(datasource);
        return tokenRepository;
    }
}
```

其中 rememberMeParameter("rememberme")方法指定了"记住我"勾选框的 name 属性，如果页面中使用了默认的 rememberme，则该方法可以省略。tokenValiditySeconds(60)方法设置了状态有效时间为 60 秒，默认为两周时间。

（6）重启项目。

登录页面如图 5-21 所示。

图 5-21 "记住我"登录页面

勾选并成功登录后，可以看到网页多了一个 RememberMe 的 Cookie 对象，如图 5-22 所示。

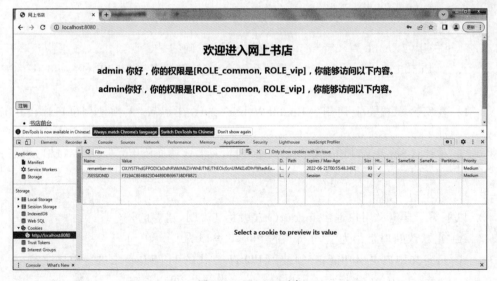

图 5-22 Cookie 对象

查看数据库表 persistent_logins，如图 5-23 所示。

可以看到 Token 信息已经成功持久化，并且浏览器也成功生成了相应的 Cookie。在 Cookie 未失效之前，无论是重开浏览器还是重启项目，用户都无须再次登录就可以访问系统资源。

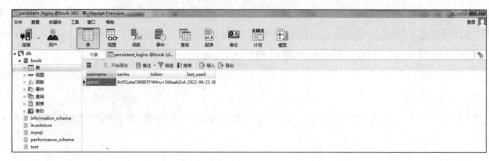

图 5-23　数据库表 persistent_logins

本章小结

本章主要讲解了 Spring Boot 的 Spring Security 安全管理。首先介绍了 Spring Security 安全框架和 Spring Boot 支持的安全管理，并体验了 Spring Boot 默认的安全管理；然后讲解了 Spring Security 自定义用户认证和授权管理；最后介绍了 Security 与前端的整合实现页面安全管理控制。希望大家通过本章的学习，能够掌握 Spring Boot 的安全管理机制，并灵活运用在实际开发中，提升项目的安全性。

在线测试

习题

一、单选题

1. 以下关于@EnableWebSecurity 注解的说法，正确的是（　　）。
   A. @EnableWebSecurity 注解是一个组合注解，开启基于 WebFluxSecurity 的安全支持
   B. 在安全配置类上使用@EnableWebSecurity 注解后，无须使用@Configuration 注解
   C. 是针对 SpringWebFlux 框架的安全支持，只需要替换使用@EnableWebFluxSecurity 注解即可
   D. 以上说法都错误

2. 以下关于自定义用户退出 logout()方法及其说明，错误的是（　　）。
   A. 它默认处理路径为"/logout"的 POST 类型请求
   B. 自定义用户退出功能，必须使用 POST 方式的 HTTP 请求进行用户注销
   C. logoutUrl()方法指定了用户退出的请求路径，可以省略
   D. 在用户退出后，用户会话信息则会默认清除

3. 以下关于基于简单加密 Token 方式的"记住我"的说法，错误的是（　　）。
   A. 基于简单加密 Token 的方式中的 Token 在指定的时间内有效
   B. 必须保证 Token 中所包含的 username、password 和 key 没有被改变
   C. 任何人获取到该"记住我"功能的 Token 后，都可以无限制进行自动登录
   D. 在 Token 有效期过后再次访问项目时，会发现又需要重新进行登录认证

4. 以下关于 Spring Boot 整合 Security 的说法,错误的是(　　)。

   A. Spring Boot 一旦引入 spring-boot-starter-security,无须配置,Spring Security 即可生效

   B. Spring Boot 整合 Security 项目启动时会在控制台 Console 中自动生成一个安全密码,每次都不一样

   C. 访问 Spring Boot 项目默认首页 index.html,无须登录

   D. 在 Spring Security 登录页面"/login"中输入错误登录信息后,会重定向到"/login?error"页面

5. 以下使用 JDBC 身份认证方式创建用户/权限表及初始化数据的说法,错误的是(　　)。

   A. 用户表中用户名必须唯一

   B. 用户表必须提供一个 tinyint 类型的字段

   C. 用户角色值是对应权限值加上"ROLE_"前缀

   D. 用户表中插入的用户密码 password 必须是对应编码器编码后的密码

二、多选题

1. 以下关于 Security 中基于持久化 Token 方式的"记住我"的说法,正确的是(　　)。

   A. 选择"记住我"并成功登录后,会把 username、随机产生的序列号、生成的 Token 进行持久化存储

   B. 当用户再次访问系统时,将重新生成一个新的 Token 替换数据库中旧的 Token

   C. 如果再次登录的 Cookie 中的 Token 不匹配,Spring Security 将删除数据库中与当前用户相关的所有 Token 记录

   D. 如果用户访问系统时没有携带 Cookie,那么将会引导用户到登录页面

2. Spring Boot 整合 Spring Security 安全框架中包含的安全管理功能包括(　　)。

   A. WebFlux Security   B. MVC Security

   C. OAuth2   D. Actuator Security

3. 以下关于 configure()方法中使用 JDBC 身份认证的方式进行自定义用户认证相关说法,正确的是(　　)。

   A. 要引入 DataSource 数据源

   B. 使用 JDBC 身份认证时,首先需要对密码进行编码设置

   C. 在定义用户查询的 SQL 语句时,必须返回用户名 username 和密码 password 两个字段信息

   D. 在定义权限查询的 SQL 语句时,必须返回用户名 username、角色 role、权限 authority 三个字段信息

4. 针对自定义用户认证,SpringSecurity 提供了多种自定义认证方式,包括(　　)。

   A. In-Memory Authentication(内存身份认证)

   B. JDBC Authentication(JDBC 身份认证)

   C. LDAP Authentication(LDAP 身份认证)

   D. UserDetailsService(身份详情服务)

5. 以下关于 Security 与 Thymeleaf 整合实现前端页面管理的相关标签及属性的说法,

错误的是(　　)。

  A. 页面顶部通过"xmlns:sec"引入了 Security 安全标签

  B. 使用 sec:authorize="! isAuthenticated()"属性判断用户是否未登录

  C. 使用 sec:authorize="hasRole('common')"属性判断用户是否有 ROLE_common 权限

  D. 使用 sec:authentication="principal.authorities"属性可以获取登录用户角色

三、判断题(对的打"√",错的打"×")

1. 自定义 WebSecurityConfigurerAdapter 类型的 Bean 组件,会同时关闭 UserDetailsService 用户信息自动配置类。　　　　　　　　　　　　　　　　　　　　　　　　　　　(　　)

2. 自定义的登录页跳转路径必须与数据处理提交路径一致。　　　　　　　　(　　)

3. rememberMeParameter()方法用于指定"记住我"勾选框的 name 属性值,可以省略。　　　　　　　　　　　　　　　　　　　　　　　　　　　　　　　　　(　　)

4. 持久化 Token 的方式比简单加密 Token 的方式相对更加安全,不会存在安全问题。　　　　　　　　　　　　　　　　　　　　　　　　　　　　　　　　　(　　)

5. Security 会默认提供一个可登录的用户信息,其中用户名为 user,密码为 root。　　　　　　　　　　　　　　　　　　　　　　　　　　　　　　　　　　(　　)

四、填空题

1. 在 Spring Security 默认登录页面中输入错误登录信息后会重定向到_____页面。

2. In-Memory Authentication(内存身份认证)是最简单的身份认证方式,主要用于_____。

3. 用户请求控制相关方法中的 permitAll()方法用于表示_____。

4. Security 控制登录的用户信息被封装在_____类对象中。

5. _____是 Security 提供的进行认证用户信息封装的接口。

五、简答题

1. 简述在 Spring Boot 项目中添加 Security 整合 Thymeleaf 进行前端页面管理依赖要注意的问题。

2. 对 WebSecurityConfigurerAdapter 类的两个主要方法进行简要介绍。

第6章

Spring Boot消息服务

本章学习目标
- 了解为什么要使用消息中间件。
- 熟悉 RabbitMQ 消息中间件的基本概念和工作原理。
- 熟悉 Spring Boot 与 RabbitMQ 的整合搭建。
- 熟悉 Spring Boot 与 RabbitMQ 整合实现常用的工作模式。

根据党的二十大报告,要增强中华文明传播力和影响力。文明因交流而多彩,文明因互鉴而丰富。只有坚持推动文明相通、文化相融,才能更好地构建人类命运共同体。

本章将针对 Spring Boot 消息服务进行讲解,并进而体会消息在文明交流中的重要性。

6.1 消息服务概述

消息队列(alignment)是一种进程间或线程间的异步通信方式。消息生产者在产生消息后,会将消息保存在消息队列中,直到消息消费者来取走它,即消息的发送者和接收者不需要同时与消息队列交互。使用消息队列可以有效实现服务的解耦,并提高系统的可靠性和可扩展性。

目前,开源的消息队列服务非常多,如 Apache ActiveMQ、RabbitMQ、Kafka、RocketMQ、ZeroMQ、Apollo 等,这些产品也就是常说的消息中间件。

1. 什么是 AMQP

AMQP(Advanced Message Queuing Protocol,高级消息队列协议)是一个线路层的协议规范,而不是 API 规范(如 JMS)。由于 AMQP 是一个线路层协议规范,因此它天然就是跨平台的,就像 SMTP、HTTP 等协议一样,只要开发者按照规范的格式发送数据,任何平台都可以通过 AMQP 进行消息交互。像目前流行的 StormMQ、RabbitMQ 等都实现了 AMQP,其实体图如图 6-1 所示。

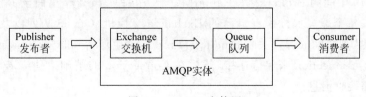

图 6-1 AMQP 实体图

AMQP 的工作过程是发布者(Publisher)发布消息(Message)，交换机根据路由规则将收到的消息分发给与该交换机绑定的队列(Queue)，AMQP 代理将消息投递给订阅了此队列的消费者或由消费者按照需求自行获取。

(1) 发布者、交换机、队列和消费者都可以有多个，同时因为 AMQP 是一个网络协议，所以这个过程中的发布者、消费者、消息代理可以分别存在于不同的设备上。

(2) 发布者发布消息时可以给消息指定各种消息属性(Message Meta-data)，有些属性可能会被消息代理(Brokers)使用，然而其他的属性则是完全不透明的，它们只能被接收消息的应用所使用。

(3) 从安全角度考虑，网络是不可靠的，而且消费者在处理消息的过程中可能会意外挂掉，这样没有处理成功的消息就会丢失。基于此原因，AMQP 模块包含了一个消息确认(Message Acknowledgements)机制：当一个消息从队列中投递给消费者后，不会立即从队列中删除，直到它收到来自消费者的确认回执(Acknowledgement)后，才完全从队列中删除。

(4) 在某些情况下，例如，当一个消息无法被成功路由(无法从交换机分发到队列)时，消息或许会被返回给发布者并被丢弃。又如，如果消息代理执行了延期操作，消息会被放入一个所谓的死信队列中。此时，消息发布者可以选择某些参数来处理这些特殊情况。

2. 什么是 RabbitMQ

通常谈到消息队列就会联想到其中的三者：生产者、消费者和消息队列。生产者将消息发送到消息队列，消费者从消息队列中获取消息进行处理。对于 RabbitMQ，它在此基础上做了一层抽象，引入了交换器 Exchange 的概念。交换器是作用于生产者和消息队列的中间桥梁，它起了一种消息路由的作用，也就是说生产者并不和消息队列直接关联，而是先发送给交换器，再由交换器路由到对应的队列。至于它是根据何种规则路由到消息队列的，就是下面需要介绍的内容。这里的生产者并没有直接将消息发送给消息队列，而是通过建立与 Exchange(交换机)的 Channel(信道)，将消息发送给 Exchange；Exchange 根据路由规则，将消息转发给指定的消息队列；消息队列储存消息，等待消费者取出消息；消费者通过建立与消息队列相连的 Channel，从消息队列中获取消息，如图 6-2 所示。

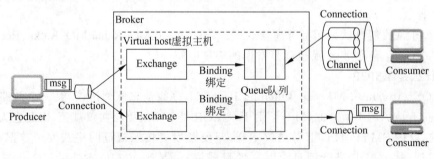

图 6-2　消息队列图

在图 6-2 中，中间的 Broker 表示 RabbitMQ 服务，每个 Broker 里面至少有一个 Virtual host 虚拟主机，每个虚拟主机中有自己的 Exchange 交换机、Queue 队列以及 Exchange 交换机与 Queue 队列之间的绑定关系 Binding。Producer(生产者)和 Consumer(消费者)通过与 Broker 建立 Connection 来保持连接，然后在 Connection 的基础上建立若干 Channel 信道，用来发送和接收消息。

(1) Connection(连接)。

每个 Producer(生产者)或 Consumer(消费者)要通过 RabbitMQ 发送和消费消息,首先就要与 RabbitMQ 建立连接,这个连接就是 Connection。Connection 是一个 TCP 长连接。

(2) Channel(信道)。

Channel 是在 Connection 的基础上建立的虚拟连接,RabbitMQ 中大部分的操作都是使用 Channel 完成的,比如声明 Queue、声明 Exchange、发布消息、消费消息等。

既然已经有了 Connection,完全可以使用 Connection 完成 Channel 的工作,为什么还要引入 Channel 这样一个虚拟连接的概念呢?因为现在的程序都是支持多线程的,如果没有 Channel,那么每个线程在访问 RabbitMQ 时都要建立一个 Connection 这样的 TCP 连接。对于操作系统来说,建立和销毁 TCP 连接是非常大的开销,在系统访问流量高峰时,会严重影响系统性能。

Channel 就是为了解决这种问题而提出的。在通常情况下,每个线程创建单独的 Channel 进行通信,每个 Channel 都有自己的 id 帮助 Broker 和客户端识别 Channel,所以 Channel 之间是完全隔离的。Connection 与 Channel 之间的关系可以这样理解,如果把 Connection 比作一条光纤电缆,那么 Channel 就相当于是电缆中的一束光纤。

(3) Virtual host(虚拟主机)。

Virtual host 是一个虚拟主机的概念,一个 Broker 中可以有多个 Virtual host,每个 Virtual host 都有一套自己的 Exchange 和 Queue,同一个 Virtual host 中的 Exchange 和 Queue 不能重名,不同的 Virtual host 中的 Exchange 和 Queue 名字可以相同。这样,不同的用户在访问同一个 RabbitMQ Broker 时,可以创建自己单独的 Virtual host,然后在自己的 Virtual host 中创建 Exchange 和 Queue,很好地做到了不同用户之间相互隔离的效果,如图 6-3 所示。

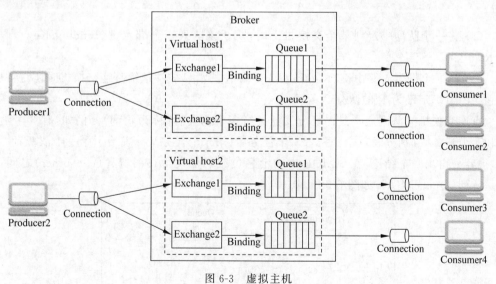

图 6-3 虚拟主机

(4) Queue(队列)。

Queue 是一个用来存放消息的队列,生产者发送的消息会被放到 Queue 中,消费者消费消息时也是从 Queue 中取走消息。

（5）Exchange（交换机）。

Exchange 是一个比较重要的概念，它是消息到达 RabbitMQ 的第一站，主要负责根据不同的分发规则将消息分发到不同的 Queue，供订阅了相关 Queue 的消费者消费到指定的消息。那 Exchange 有哪些分发消息的规则呢？这就要说到 Exchange 的 4 种类型了，即 direct、fanout、topic 和 headers。

6.2　Exchange 策略

在介绍 Exchange 的 4 种类型之前，先来了解另一个比较重要的概念 Routing Key，即路由键。当创建好 Exchange 和 Queue 后，需要使用 Routing Key（通常叫作 Binding Key）将它们绑定起来，Producer 在向 Exchange 发送一条消息时，必须指定一个 Routing Key。Exchange 接收到这条消息后，会解析 Routing Key，然后根据 Exchange 和 Queue 的绑定规则，将消息分发到符合规则的 Queue 中，如图 6-4 所示。

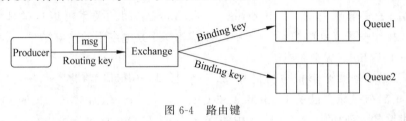

图 6-4　路由键

1. 直连交换机

直连（direct）交换机根据消息携带的路由键（Routing Key）将消息投递给对应绑定键的队列，用来处理消息的单播路由（route）（尽管它也可以处理多播路由）。下面介绍它的工作流程。

（1）将一个队列绑定到某个交换机上时，赋予该队列一个绑定键（Binding Key），假设为 R。

（2）当一个携带着路由键（Routing Key）为 R 的消息被发送给直连交换机时，交换机会将它路由给绑定键为 R 的队列。

直连交换机的队列通常是循环分发任务给多个消费者（称为轮询）。例如，有 3 个消费者和 4 个任务，分别分发每个消费者一个任务后，第 4 个任务又分发给了第一个消费者。综上，很容易得出一个结论，在 AMQP 中的消息负载均衡发生在消费者（consumer）之间，而不是队列（Queue）之间，如图 6-5 所示。

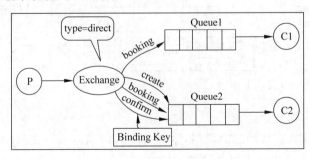

图 6-5　直连交换机

当生产者（P）发送消息且 Routing Key＝booking 时，将消息传送给 Exchange，Exchange 获取到生产者发送过来的消息后，会根据自身的规则匹配相应的 Queue，这时发现 Queue1 和 Queue2 都符合，就会将消息传送给这两个队列。

如果以 Routing Key＝create 和 Routing Key＝confirm 发送消息时，消息只会被推送到 Queue2 队列中，其他 Routing Key 的消息将会被丢弃。

2．扇形交换机

扇形（Fanout）交换机将消息路由给绑定到它身上的所有队列，而不理会绑定的路由键。如果 N 个队列绑定到某个扇形交换机上，当有消息发送给此扇形交换机时，交换机会将消息的复制分别发送给这 N 个队列。扇形交换机用来处理消息的广播路由（route），如图 6-6 所示。

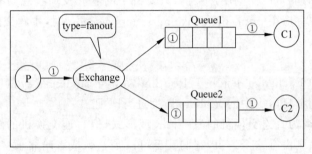

图 6-6　扇形交换机

因为扇形交换机投递消息时复制到所有绑定到它的队列，所以它的应用案例都极其相似：

（1）大规模多用户在线（MMO）游戏可以使用它来处理排行榜更新等全局事件。

（2）体育新闻网站可以用它来近乎实时地将比分更新分发给移动客户端。

（3）分发系统使用它来广播各种状态和配置更新。

（4）它在群聊时被用来分发消息给参与群聊的用户。

如图 6-6 所示，生产者（P）生产消息 1 时将消息 1 推送到 Exchange，由于 Exchange type＝fanout，这时会遵循 fanout 的规则将消息推送到所有与它绑定的 Queue，也就是图 6-6 中的两个 Queue，最后两个消费者消费。

3．主题交换机

前面提到的 direct 规则是严格意义上的匹配，即 Routing Key 必须与 Binding Key 相匹配时才将消息传送给 Queue。

而主题（topic）交换机的路由规则是一种模糊匹配，通过通配符满足一部分规则就可以传送。

topic 规则的约定如下：

（1）Binding Key 中可以存在两种特殊字符"＊"与"＃"，用于模糊匹配，其中"＊"用于匹配一个单词，"＃"用于匹配多个单词（可以是零个）。

（2）Routing Key 为一个句点号"．"分隔的字符串，如"stock.usd.nyse""nyse.vmw""quick.orange.rabbit"。Binding Key 与 Routing Key 一样，也是句点号"．"分隔的字符串。

当生产者发送消息 Routing Key＝F.C.E 时，这时只满足 Queue1，所以会被路由到 Queue1 中；如果 Routing Key＝A.C.E，会同时路由到 Queue1 和 Queue2 中；如果

Routing Key＝A. F. B 时，只会发送一条消息到 Queue2 中，如图 6-7 所示。

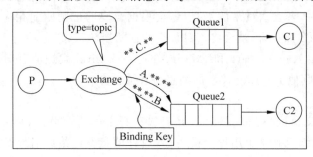

图 6-7　主题交换机

主题交换机拥有非常广泛的用户案例，只要一个问题涉及那些想要有针对性地选择接收消息的多消费者/多应用（multiple consumers/applications）时，主题交换机就可以被列入考虑范围。

4. 头交换机

headers 类型的 Exchange 不依赖于 Routing Key 与 Binding Key 的匹配规则来路由消息，而是根据发送的消息内容中的 headers 属性进行匹配。

头交换机可以视为直连交换机的另一种表现形式，但直连交换机的路由键必须是一个字符串，而头属性值则没有这个约束，它们甚至可以是整数或者哈希值（字典）等，灵活性更强。头交换机的工作流程如下：

（1）绑定一个队列到头交换机上，同时绑定多个用于匹配的头（header）。

（2）传来的消息会携带 header，以及一个"x-match"参数。当"x-match"设置为"any"时，消息头的任意值被匹配就可以满足条件；当"x-match"设置为"all"时，就需要消息头的所有值都匹配成功。

6.3　消息的各种机制

1. 消息确认机制

在实际应用中，可能会发生消费者收到 Queue 中的消息，但没有处理完成就宕机（或出现其他意外）的情况，这种情况下就可能会导致消息丢失。为了避免这种情况发生，可以要求消费者在消费完消息后发送一个回执给 RabbitMQ，RabbitMQ 收到消息回执（message acknowledgment）后才将该消息从 Queue 中移除；如果 RabbitMQ 没有收到回执并检测到消费者的 RabbitMQ 连接断开，则 RabbitMQ 会将该消息发送给其他消费者（如果存在多个消费者）进行处理。这里不存在 Timeout 概念，一个消费者处理消息时间再长也不会导致该消息被发送给其他消费者，除非它的 RabbitMQ 连接断开。但这里会产生另外一个问题，如果开发人员在处理完业务逻辑后，忘记发送回执给 RabbitMQ，这将会导致严重的问题，Queue 中堆积的消息会越来越多，消费者重启后会重复消费这些消息并重复执行业务逻辑。如果采用 no-ack 的方式进行确认，也就是说，每次 Consumer 接收到数据后无论是否处理完成，RabbitMQ 都会立即把这个 Message 标记为完成，然后从 Queue 中删除。

2. 消息持久化机制

如果希望即使在 RabbitMQ 服务重启的情况下也不会丢失消息，可以将 Queue 与

Message 都设置为可持久化的(durable),这样可以保证绝大部分情况下的 RabbitMQ 消息不会丢失,但依然避免不了小概率丢失事件的发生(如 RabbitMQ 服务器已经接收到生产者的消息,但还没来得及持久化该消息时,RabbitMQ 服务器就断电了)。如果需要对这种小概率事件进行管理,就要用到事务。

3. 事务

对事务的支持是 AMQP 协议的一个重要特性。假设当生产者将一个持久化消息发送给服务器时,因为 consume 命令本身没有任何 Response 返回,所以即使服务器崩溃而没有持久化该消息,生产者也无法获知该消息已经丢失。如果此时使用事务,即通过 txSelect() 开启一个事务,然后发送消息给服务器并通过 txCommit() 提交该事务,这样就可以保证该消息在 txCommit() 提交后一定会被持久化。如果 txCommit() 还未提交即服务器崩溃,则该消息不会被服务器接收。此外,Rabbit MQ 还提供了 txRollback() 命令用于回滚某一个事务。

4. 消息分发机制

在应用程序使用消息系统时,一般情况下生产者往队列里插入数据的速度是比较快的,但是消费者消费数据往往涉及一些业务逻辑处理,从而导致速度跟不上生产者生产数据。因此如果一个生产者对应一个消费者,容易导致很多消息堆积在队列里,这时就需要使用工作队列。一个队列有多个消费者同时消费数据。工作队列有两种分发数据的方式:轮询分发(round-robin)和公平分发(fair dispatch)。轮询分发是指队列给每个消费者发送数量相同的数据。公平分发是指消费者设置每次从队列中取一条数据,并且消费完后手动应答,继续从队列取下一个数据。

(1) 轮询分发。

如果工作队列中有两个消费者,两个消费者得到的数据量是相同的,并不会因为两个消费者处理数据速度不同而取得不同数量的数据。但是这种分发方式存在一些隐患,消费者虽然得到了消息,但是如果消费者没能成功处理业务逻辑,在 RabbitMQ 中也不会存在这条消息,就会出现消息丢失且业务逻辑没能成功处理的情况。

(2) 公平分发。

消费者设置每次从队列里取一条数据,手动回复并继续取下一条数据,并且关闭自动回复机制。这样队列就会公平地给每个消息费者发送数据,消费一条后再发第二条,而且可以在管理界面中看到数据是一条条随着消费者消费完而减少的,并不是一下子全部分发完。采用公平分发的方式就不会出现消息丢失且业务逻辑没能成功处理的情况。

6.4 安装 RabbitMQ

6.4.1 什么是 RabbitMQ

RabbitMQ 是一个由 Erlang 语言开发的 AMQP 的开源实现。它是应用层协议的一个开放标准,为面向消息的中间件设计。基于此协议的客户端与消息中间件可传递消息,并不受产品、开发语言等条件的限制。RabbitMQ 最初起源于金融系统,用于在分布式系统中存储转发消息,在易用性、扩展性、高可用性等方面表现出色。

RabbitMQ 的具体特点如下:

（1）可靠性（reliability）。RabbitMQ 使用一些机制来保证可靠性，如持久化、传输确认、发布确认等。

（2）灵活的路由（route）。在消息进入队列之前，通过 Exchange 来路由消息。对于典型的路由功能，RabbitMQ 已经提供了一些内置的 Exchange 来实现。针对更复杂的路由功能，可以将多个 Exchange 绑定在一起，也可以通过插件机制实现自己的 Exchange。

（3）消息集群（message cluster）。多个 RabbitMQ 服务器可以组成一个集群，形成一个逻辑 Broker。

（4）高可用队列（highly available queues）。RabbitMQ 可以在集群中的机器上进行镜像，使得在部分节点出问题的情况下队列仍然可用。

（5）多种协议（multi-protocol）。RabbitMQ 支持多种消息队列协议，如 STOMP、MQTT 等。

（6）多语言客户端（multilingual client）。RabbitMQ 几乎支持所有常用语言，如 Java、.NET、Ruby 等。

（7）管理界面（management UI）。RabbitMQ 提供了一个易用的用户界面，使得用户可以监控和管理消息 Broker 的许多方面。

（8）跟踪机制（tracking mechanism）。RabbitMQ 提供了消息跟踪机制，使用者可以查出消息、异常的原因。

（9）插件机制（plug-in mechanism）。RabbitMQ 提供了许多插件，从多方面进行扩展，也可以编写自己的插件。

6.4.2 RabbitMQ 安装过程

在安装 RabbitMQ 前需要先安装部署 Erlang 环境，并且要注意 Erlang 的安装版本与 RabbitMQ 的版本号相对应，如图 6-8 所示。

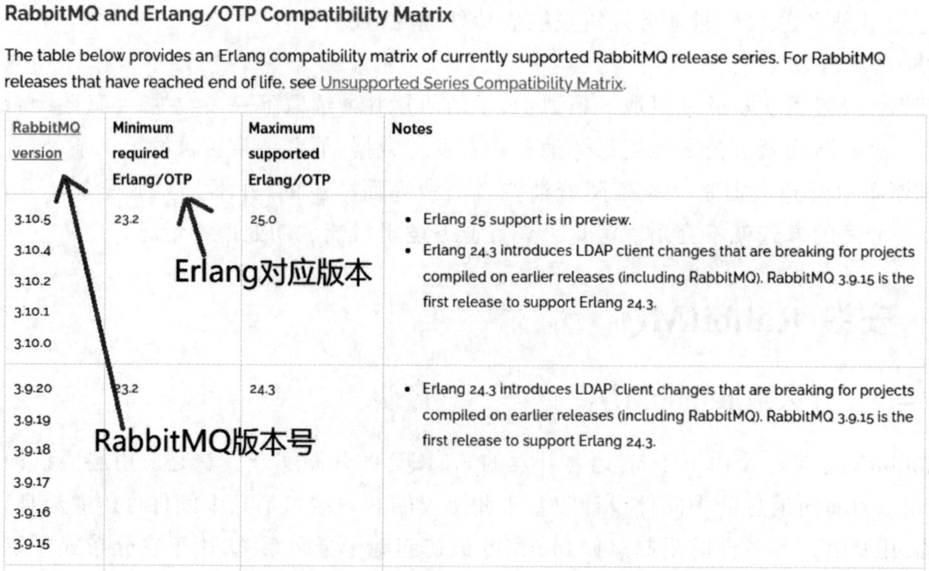

RabbitMQ and Erlang/OTP Compatibility Matrix

The table below provides an Erlang compatibility matrix of currently supported RabbitMQ release series. For RabbitMQ releases that have reached end of life, see Unsupported Series Compatibility Matrix.

RabbitMQ version	Minimum required Erlang/OTP	Maximum supported Erlang/OTP	Notes
3.10.5 3.10.4 3.10.2 3.10.1 3.10.0	23.2	25.0	• Erlang 25 support is in preview. • Erlang 24.3 introduces LDAP client changes that are breaking for projects compiled on earlier releases (including RabbitMQ). RabbitMQ 3.9.15 is the first release to support Erlang 24.3.
3.9.20 3.9.19 3.9.18 3.9.17 3.9.16 3.9.15	23.2	24.3	• Erlang 24.3 introduces LDAP client changes that are breaking for projects compiled on earlier releases (including RabbitMQ). RabbitMQ 3.9.15 is the first release to support Erlang 24.3.

图 6-8　Erlang 和 RabbitMQ 相对应的版本号

1. 安装 Erlang

Erlang 是一个结构化动态类型编程语言，内建并行计算支持。它最初是由爱立信专门为通信应用设计的，比如控制交换机或变换协议等，因此非常适合于构建分布式实时软并行计算系统。使用 Erlang 编写出的应用在运行时通常由成千上万个轻量级进程组成，并通过消息传递相互通信。进程间上下文切换对于 Erlang 来说仅仅只是一两个环节，比起 C 程序的线程切换要高效得多。

Erlang 的安装步骤如下。

(1) 下载地址为 https://www.erlang.org/downloads，选择合适的版本，这里选择 21.2 版本，如图 6-9 所示。

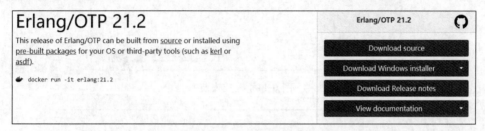

图 6-9　选择 Erlang 安装版本

(2) 运行 otp_win64_21.2.exe，选择安装的位置，如图 6-10 所示。依次单击 Next→Install 按钮完成安装，如图 6-11 所示。

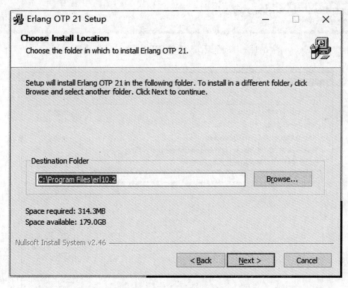

图 6-10　安装位置

(3) 设置环境变量 ERLANG_HOME。右击我的电脑并右击属性，选择高级系统设置→环境变量→系统变量命令，单击"新建"按钮，在弹出的新建系统变量选项卡中填入变量名 ERLANG_HOME 和变量值 C:\Program Files\erl10.2(C:\Program Files\erl10.2 即为安装目录)，单击"确定"按钮完成设置环境变量，如图 6-12 所示。

(4) 配置完上面的环境变量后，找到系统变量中的 PATH。单击"编辑"按钮，在末尾追加％ERLANG_HOME％\bin。单击"确定"按钮，完成编辑环境变量，如图 6-13 所示。

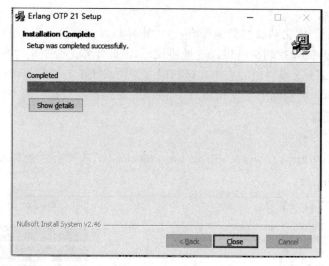

图 6-11 完成安装

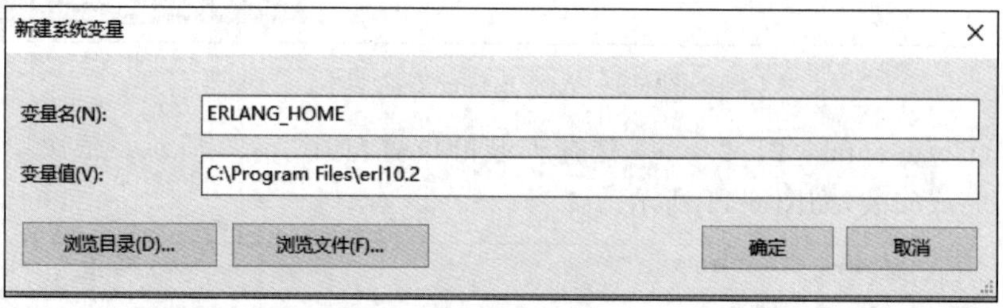

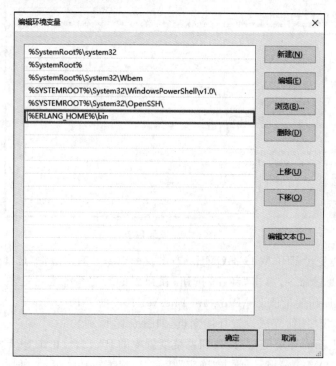

图 6-12 新建环境变量 ERLANG_HOME

图 6-13 设置 path 路径

（5）安装完成后，按下 Win＋R 组合键启动 cmd，查看是否配置成功。打开命令行，输入 erl。当提示版本信息为 Eshell V10.2（abort with ^G）时，说明 Erlang 安装成功，如图 6-14 所示。

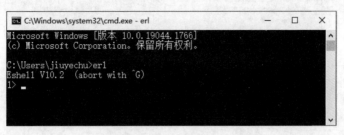

图 6-14 验证 Erlang 安装成功

2. 安装 RabbitMQ

（1）下载地址为 https://www.rabbitmq.com/install-windows.html，选择和 Erlang 匹配的版本进行下载，这里选择 rabbitmq-server-3.7.9.exe。

（2）运行 rabbitmq-server-3.7.9.exe，安装 RabbitMQ，如图 6-15 所示。此处需要注意的是，如果要自定义安装路径，路径中最好不要存在中文，否则会出现错误。

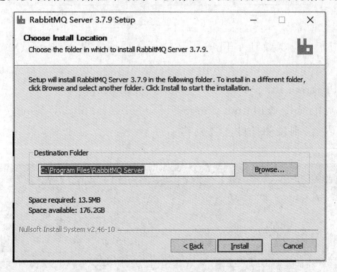

图 6-15 设置安装路径

（3）设置环境变量，方法与设置 Erlang 的环境变量方法相同。这里新建变量名为 RABBITMQ_SERVER，变量值为 C:\Program Files\RabbitMQ Server\rabbitmq_server-3.7.9，如图 6-16 所示。

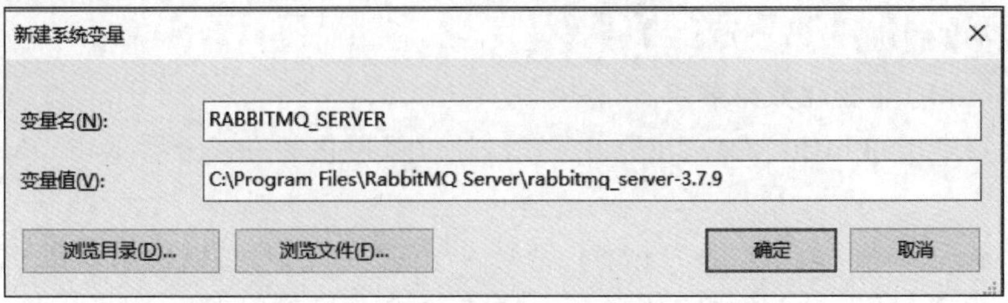

图 6-16 新建环境变量 RABBITMQ_SERVER

（4）修改 path 路径，追加值为%RABBITMQ_SERVER%\sbin，如图 6-17 所示。

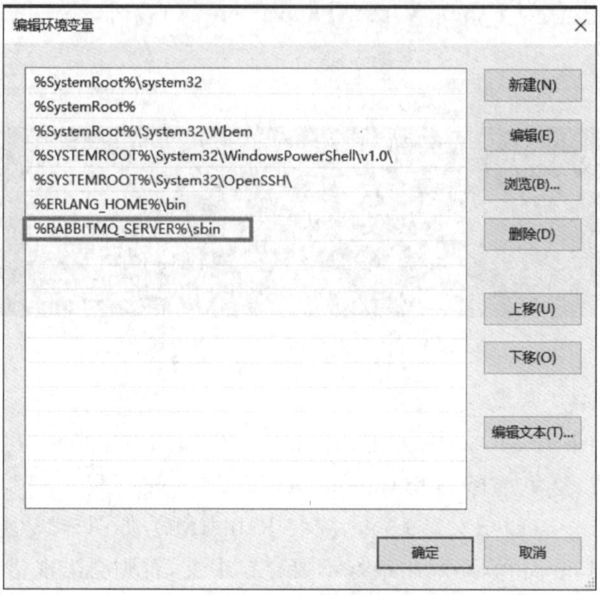

图 6-17　设置 path 路径

（5）安装完成后，需要激活 rabbitmq_management。进入 RabbitMQ 安装目录 C:\Program Files\RabbitMQ Server\rabbitmq_server-3.7.9\sbin，命令行先切换到 C 盘，然后执行命令 cd C:\Program Files\RabbitMQ Server\rabbitmq_server-3.7.9\sbin。将目录切换到指定目录后，运行命令 rabbitmq-plugins.bat enable rabbitmq_management。如果出现如图 6-18 所示的内容，则说明插件安装成功。

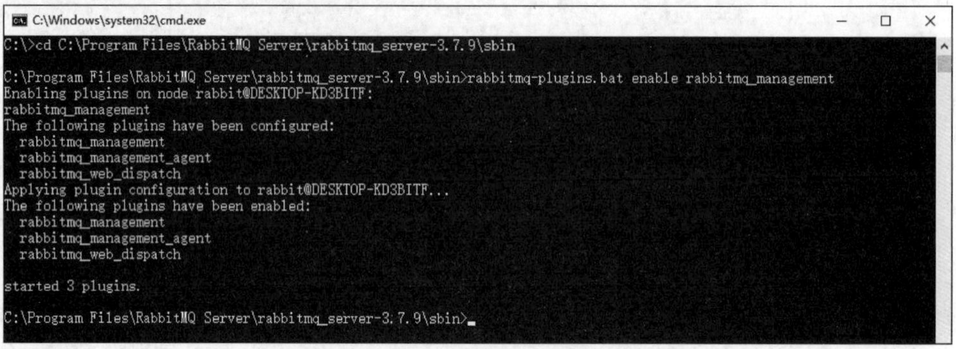

图 6-18　验证插件安装成功

（6）在命令行输入 rabbitmqctl status，如果出现如图 6-19 所示的内容，则证明 RabbitMQ 安装成功。

3. RabbitMQ 可视化效果展示

在浏览器上访问 http://localhost:15672/，通过可视化的方式查看 RabbitMQ，效果如图 6-20 所示。

系统默认会提供一个用户名为 guest、密码为 guest 的用户，可以使用该账号进行登录。登录成功后会进入 RabbitMQ 可视化管理页面的首页，如图 6-21 所示。

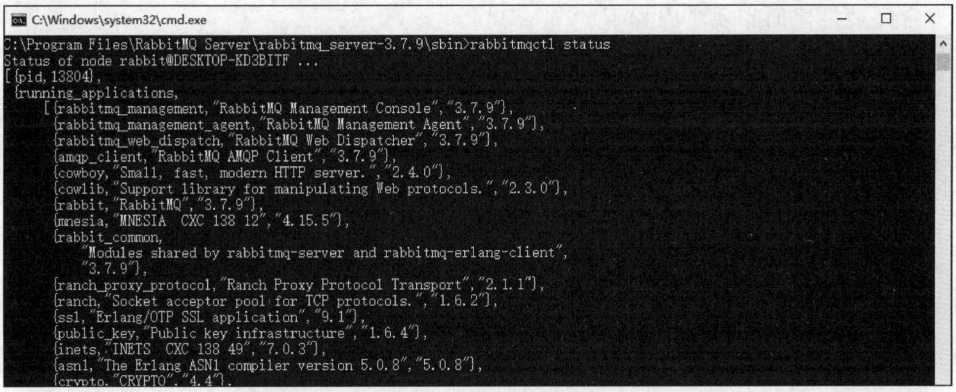

图 6-19　验证 RabbitMQ 安装成功

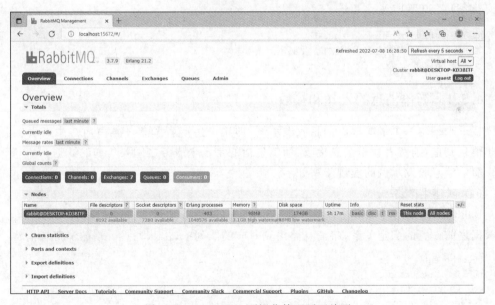

图 6-20　以可视化的方式查看 RabbitMQ 效果图

图 6-21　RabbitMQ 可视化管理页面首页

6.5　Spring Boot 整合 RabbitMQ

完成 RabbitMQ 的安装后，就可以开始对 Spring Boot 整合 RabbitMQ 实现消息服务需要的整合环境进行搭建。

整合环境搭建的实现步骤如下。

（1）创建项目并添加相应依赖。

新建一个名为 springboot0601 的项目，依次选择 Web 依赖和 Messaging 模块中的 RabbitMQ 依赖。在 pom.xml 文件中，spring-boot-starter-amqp 里面的 amqp 指的是高级消息队列协议，而 RabbitMQ 就是 AMQP 协议的一种实现中间件。

```xml
<dependencies>
    <!-- amqp -->
    <dependency>
        <groupId>org.springframework.boot</groupId>
        <artifactId>spring-boot-starter-amqp</artifactId>
    </dependency>
    <!-- web -->
    <dependency>
        <groupId>org.springframework.boot</groupId>
        <artifactId>spring-boot-starter-web</artifactId>
    </dependency>
</dependencies>
```

（2）编写配置文件。

在 application.properties 文件中配置 RabbitMQ 的地址和密码，这里使用默认的虚拟主机，如果已经创建则可以通过配置指定对应的虚拟主机。

```
# RabbitMQ 服务的地址
spring.rabbitmq.host=localhost
spring.rabbitmq.port=5672
spring.rabbitmq.username=guest
spring.rabbitmq.password=guest
# RabbitMQ 服务创建的虚拟主机(非必需)
# spring.rabbitmq.virtual-host=/
```

6.5.1 简单消息的发送和接收

视频讲解

接下来编写一个最简单的消息收发示例，即单一生产者和单一消费者。生产者实际上就是消息制造，消费者就是消息接收，而消息指的是各个服务之间要传递的数据。

简单消息收发示例的实现步骤如下。

（1）创建生产者。

RabbitMQ 生产者直接注入 RabbitTemplate，调用方法传入交换机名称及路由键。它会根据之前设置的绑定关系，将消息路由到对应的队列中，由队列另一端的消费者消费。

```java
@Component
public class Producer {
    @Autowired
    private RabbitTemplate rabbitTemplate;
    public void produce() {
        String message;
        for (int i = 1; i < 11; i++) {
            message = "第 i 位老师说疫情期间注意防护";
            rabbitTemplate.convertAndSend("notice_queue", message);
        }
    }
}
```

(2) 创建测试类发送消息。

```
@SpringBootTest
class Springboot0601ApplicationTests {
    @Autowired
    Producer producer;
    @Test
    public void amqpTest() throws InterruptedException {
        //生产者发送消息
        producer.produce();
    }
}
```

访问 RabbitMQ Web 页面,效果如图 6-22 所示,此时消息队列中存在 10 条待消费的消息。单击队列名 notice_queue,进入队列的消息信息页面,如图 6-23 所示。

图 6-22　查看队列信息

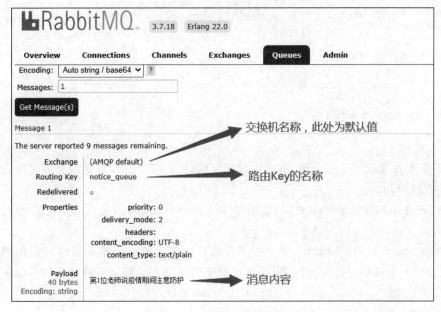

图 6-23　查看队列详细消息

（3）创建消费者。

RabbitMQ 消费者使用@RabbitListener 注解指定对应的队列即可消费消息。需要注意的是，在消费者中需要处理一下异常，如果不处理会导致消费者无法继续消费。

```
@Component
public class Consumer {
    @RabbitHandler
    @RabbitListener(queuesToDeclare = @Queue("notice_queue"))
    public void receivel(String message) {
        System.out.println("receivel 同学收到通知： " + message);
    }
    @RabbitHandler
    @RabbitListener(queuesToDeclare = @Queue("notice_queue"))
    public void receive2(String message) {
        System.out.println("receive2 同学收到通知： " + message);
    }
}
```

Consumer 消费者通过@RabbitListener 注解创建侦听器端点，并绑定 notice_queue 队列。

① @RabbitListener 注解提供了@QueueBinding、@Queue、@Exchange 等对象，通过这个组合注解配置交换机、绑定路由和配置监听功能等。设置该消费者监听哪个队列，即队列中一有消息就拿来消费。

② @RabbitHandler 注解设置自动确认，即从队列中拿到消息后就算确认消费了，不管消息是不是需要的消息，也不管之后的逻辑处理。

（4）消费队列中的消息。

```
@SpringBootTest
class Springboot0601ApplicationTests {
    @Autowired
    Producer producer;
    @Test
    public void amqpTest() throws InterruptedException {
        //生产者发送消息
        //producer.produce();
         //让消息飞一会儿
            Thread.sleep(2000);
    }
}
```

注销生产者发送消息的语句，再次运行此方法，可以在控制台看到消费者已经在接收消息，如图 6-24 所示。

再次访问 RabbitMQ Web 页面，效果如图 6-25 所示，此时消息队列中已经没有任何待消费的消息。

从控制台输出可以明显看出，多个消费端采用轮询的策略。轮询策略只适用于短时间内，可以设定一个类似过期时间的属性，超过这个时间没有消费过消息就重新开始轮询。

可以通过设置 channel.basicQos(1)阻止 RabbitMQ 将消息平均分配，这时会优先发给不忙的消费者，如果当前消费者忙则会发送给下一个消费者。

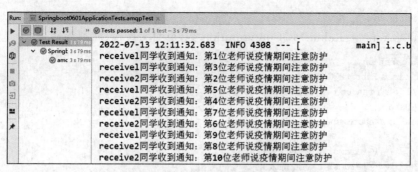

图 6-24　控制台信息

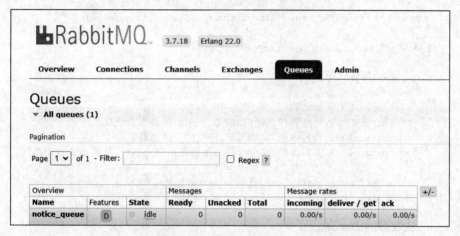

图 6-25　查看消息队列信息

6.5.2　发布订阅模型

假设有一个订单系统，用户进行下单支付，系统在用户下单成功后进行业务处理，并消息通知用户相关信息。例如，通过邮件、短信、QQ 等方式进行消息推送支付成功信息，如图 6-26 所示。

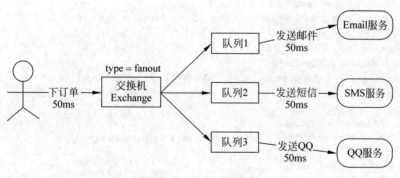

图 6-26　发布订阅模型

发布订阅的实现步骤如下：

（1）创建生产者项目并添加相应依赖。

新建一个项目 springboot0602，并添加相应的依赖。新建项目与上述项目是一样的流程,这里不再展示。

(2) 编写 Fanout_RabbitMQConfiguration 类,声明交换机、队列等信息。

```java
@Configuration
public class RabbitMQConfiguration {
    //1.声明 fanout 广播模式的交换机
    @Bean
    public FanoutExchange getExchange(){
        /**
         * @params1 : 交换机名称
         * @params2 : 是否持久化
         * @params4 : 是否自动删除
         */
        return new FanoutExchange("fanout_order_exchange",true,false);
    }
    //2.声明三个队列: emailQueue、smsQueue、qqQueue
    @Bean
    public Queue getEmailQueue(){
        /**
         * @params1 : 队列名称
         * @params2 : 队列是否持久化(如果是,则重启服务不会丢失)
         * @params3 : 是否独占队列(如果是,则仅限于此连接)
         * @params4 : 是否自动删除(当最后一条消息消费完毕时,队列是否自动删除)
         */
        return new Queue("email_fanout_Queue",true,false,false);
    }
    @Bean
    public Queue getSMSQueue(){
        return new Queue("sms_fanout_Queue",true,false,false);
    }
    @Bean
    public Queue getQqQueue(){
        return new Queue("qq_fanout_Queue",true,false,false);
    }
    //3.绑定交换机与队列的关系
    @Bean
    public Binding getEmailBinding(){
        return BindingBuilder.bind(getEmailQueue()).to(getExchange());
    }
    @Bean
    public Binding getSMSBinding(){
        return BindingBuilder.bind(getSMSQueue()).to(getExchange());
    }
    @Bean
    public Binding getQQBinding(){
        return BindingBuilder.bind(getQqQueue()).to(getExchange());
    }
}
```

这里定义一个配置类,其主要内容如下:

① 设置队列

`return new Queue(name,durable,exclusive,autoDelete)`

- name:队列名称,后续用于消费者获取消息。

- durable：是否持久化，即是否会被存储到磁盘上。当消息代理重启时仍存在，默认为 true。
- exclusive：只能被当前创建的连接使用，当连接关闭队列后立即删除，默认为 false。
- autoDelete：是否自动删除，当没有生产者或消费者使用此队列时自动删除，默认为 false。

② 设置交换机

return new DirectExchange(name,durable,autoDelete)

- name：交换机名称，后续用于消息发布者发布消息。
- durable：是否持久化，即是否会被存储到磁盘上。当消息代理重启时仍存在，默认为 true。
- autoDelete：是否自动删除，当没有生产者或队列使用该交换机时自动删除，默认为 false。

③ 绑定交换机和队列

return BindingBuilder.bind(Queue()).to(Exchange()).with(routingkey)

- Queue()：需要绑定的队列。
- Exchange()：需要绑定的交换机。
- routingKey：设置路由键，用于交换机将消息发送到队列中。

fanoutexchange 不需要设置路由键，只需要将队列和交换机绑定起来即可。在定义好这些配置后，下面就可以开始发送数据了。

(3) 创建订单服务，模拟下单。

```
@Service
public class OrderService {
    @Autowired
    private RabbitTemplate template;
    /**
     * 模拟用户创建订单
     * @param userId 客户 ID
     * @param productId 产品 ID
     * @param num 数量
     */
    public void createOrder(String userId, String productId, int num){
        //1.根据商品 ID 查询库存是否充足
        //2.生成订单
        String orderId = UUID.randomUUID().toString();
        System.out.println("订单生成成功…");
        //3.将订单 ID 封装成 MQ 消息，并投递到交换机
        /** @params1：交换机名称
         * @params2：路由 Key/队列名称
         * @params3：消息内容
         */
        template.convertAndSend("fanout_order_exchange","",orderId);
    }
}
```

（4）对测试类进行测试。

```
@SpringBootTest
class Springboot0602ApplicationTests {
    @Autowired
    private OrderService orderService;
    @Test
    void contextLoads() {
        orderService.createOrder("1001","96",1);
    }
}
```

运行测试后可以查看控制台信息，如图 6-27 所示。然后再次访问 RabbitMQ Web 页面并登录图形化界面，查看队列等信息，如图 6-28 所示。

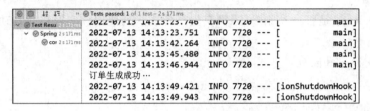

图 6-27　订单生成成功信息

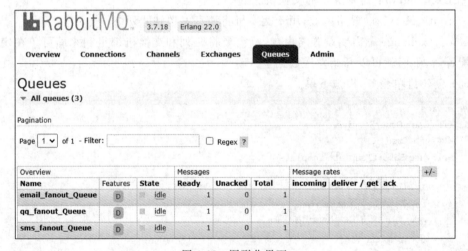

图 6-28　图形化界面

由图 6-28 可见，交换机和队列等信息已创建完毕。由于采用 fanout 广播模式的交换机，因此绑定该交换机的所有队列，消息被相应投递到各个队列。

（5）创建消费者项目并添加相应依赖。

新建一个项目 springboot0603，与项目 springboot0602 流程相同，这里不再展示。

（6）配置 application.properties 文件。

由于生产者和消费者是两个项目，生产者项目配置的端口号是 8080，因此消费者需要配置不同的端口号，否则会显示端口已被占用的异常。

```
# RabbitMQ 服务的地址
spring.rabbitmq.host = localhost
spring.rabbitmq.port = 5672
```

```
spring.rabbitmq.username = guest
spring.rabbitmq.password = guest
#消费者的端口号
server.port = 8082
```

（7）新建消费者。

新建 3 个消费者，分别监听生产者项目中声明的 3 个队列。QQ、SMS 的消费者类与 Email 消费者类一致，这里不再展示。唯一不同的是，监听队列需要相应修改为正确的名字。

```
@Component
@RabbitListener(queues = {"email_fanout_Queue"})
    //监听 email_fanout_Queue 队列
public class EmailConsumer {
    //接收消息
    @RabbitHandler
    public void receiveMess(String message) {
        System.out.println("EmailConsumer 接收到订单消息————>" + message);
    }
}
```

（8）启动主程序。

运行消费者类，结果如图 6-29 所示。由图 6-29 可见，绑定了 FanoutExchange 的队列都获得了消息，消费者也都消费成功。

图 6-29　消费者控制台

6.5.3　会员注册模型

视频讲解

假设在实现了会员注册后发送通知，消息接收者在接收到消息后进行保存会员操作，系统发送注册成功邮件，如图 6-30 所示。

图 6-30　会员注册模型

会员注册的实现步骤如下：

（1）创建项目并添加相应依赖。

新建一个项目 springboot0604，并添加相应的依赖。由于需要保存用户信息，所以需要添加 JPA 依赖，这里不再展示。

(2) 配置 application.properties 文件。

在 application.properties 文件中,配置 RabbitMQ 的相关消息和数据库连接信息。

```
# RabbitMQ 服务的地址
spring.rabbitmq.host = localhost
spring.rabbitmq.port = 5672
spring.rabbitmq.username = guest
spring.rabbitmq.password = guest
# 配置数据库连接信息
spring.datasource.url = jdbc:mysql://localhost:3306/book
spring.datasource.username = root
spring.datasource.password = 123456
spring.datasource.driver-class-name = com.mysql.jdbc.Driver
# 配置 JPA 信息
spring.data.jpa.repositories.enabled = true
spring.jpa.database = mysql
spring.jpa.generate-ddl = false
spring.jpa.hibernate.ddl-auto = update
spring.jpa.show-sql = true
```

(3) 配置数据库相关操作类。

这里采用 Spring Data JPA 方式操作数据库,首先创建一张 user 表,接着创建该类对应的实体类 User。

```
@Entity
@Table(name = "user")
@Data
public class User implements Serializable {
    @Id
    private String id;
    private String username, password, email;
}
```

在 repository 包下创建 UserRepository 接口,该接口继承了 JpaRepository,默认实现了增、删、查、改功能。

```
public interface UserRepository extends JpaRepository<User,String> {
}
```

(4) 新建常量类 Constants。

该类主要存放 RabbitMQ 队列名称、交换机,以及两者之间的路由键。

```
public class Constants {
    //消息队列-topic 交换机名称
    public static final String MEMBER_TOPIC_EXCHANGE_NAME = "rabbit_mq_topic_exchange_name";
    //消息队列-注册会员-队列名称
    public static final String MEMBER_REGISTER_QUEUE_NAME = "rabbit_mq_member_register_queue_name";
    //消息队列-注册会员-队列路由键
    public static final String MEMBER_REGISTER_QUEUE_ROUTE_KEY = "register.*";
    //消息队列-发送邮件-队列名称
    public static final String MEMBER_SEND_MAIL_QUEUE_NAME = "rabbit_mq_member_send_mail_queue_name";
    //消息队列-发送邮件-队列路由键
```

```java
    public static final String MEMBER_SEND_MAIL_QUEUE_ROUTE_KEY = "register.#";
    //消息队列-topic交换机-路由键
    public static final String MEMBER_TOPIC_EXCHANGE_ROUTE_KEY = "register.member";
}
```

(5) 新建 RabbitMQ 相关配置类。

该类主要负责创建一个 Topic Exchange 通配符交换机,实现一个注册会员队列及与交换机的绑定操作、一个发送邮件队列及与交换机进行绑定的操作,并输出各个对象实例的创建状态。

```java
@Configuration
public class UserRegistrtRabbitMQConfiguration {
    /**
     * 创建通配符交换机实例
     * @return 通配符交换机实例
     */
    @Bean
    public TopicExchange topicExchange() {
        TopicExchange topicExchange = new TopicExchange(Constants.MEMBER_TOPIC_EXCHANGE_NAME);
        System.out.println("【【【会员注册通配符交换机实例创建成功】】】");
        return topicExchange;
    }

    /**
     * 创建会员注册队列实例,并持久化
     * @return 会员注册队列实例
     */
    @Bean
    public Queue memberRegisterQueue() {
        Queue memberRegisterQueue = new Queue(Constants.MEMBER_REGISTER_QUEUE_NAME, true);
        System.out.println("【【【会员注册队列实例创建成功】】】");
        return memberRegisterQueue;
    }

    /**
     * 创建会员发送邮件队列实例,并持久化
     * @return 会员发送邮件队列实例
     */
    @Bean
    public Queue memberSendMailQueue() {
        Queue memberRegisterQueue = new Queue(Constants.MEMBER_SEND_MAIL_QUEUE_NAME, true);
        System.out.println("【【【会员发送邮件队列实例创建成功】】】");
        return memberRegisterQueue;
    }

    /**
     * 绑定会员注册队列到交换机
     * @return 绑定对象
     */
    @Bean
    public Binding memberRegisterBinding() {
        Binding binding = BindingBuilder.bind(memberRegisterQueue()).to(topicExchange()).with(Constants.MEMBER_REGISTER_QUEUE_ROUTE_KEY);
        System.out.println("【【【会员注册队列与交换机绑定成功】】】");
```

```java
        return binding;
    }
    /**
     * 绑定会员发送邮件队列到交换机
     * @return 绑定对象
     */
    @Bean
    public Binding memberSendMailBinding() {
        Binding binding = BindingBuilder.bind(memberSendMailQueue()).to(topicExchange()).with(Constants.MEMBER_SEND_MAIL_QUEUE_ROUTE_KEY);
        System.out.println("【【【会员发送邮件队列与交换机绑定成功】】】");
        return binding;
    }
}
```

(6) 新建消息发送者 MemberRegisterSender。

该类注入了 RabbitTemplate，使用 RabbitTemplate 提供的 convertAndSend 方法进行消息的发送，传入定义好的交换机、路由键和待发送的消息对象。

```java
@Component
public class MemberRegisterSender {
    @Autowired
    private RabbitTemplate rabbitTemplate;
    /**
     * 发送会员注册通知消息
     * @param message 消息内容
     */
    public void sendMessage(User message) throws Exception {
        rabbitTemplate.convertAndSend(Constants.MEMBER_TOPIC_EXCHANGE_NAME, Constants.MEMBER_TOPIC_EXCHANGE_ROUTE_KEY, message);
    }
}
```

(7) 新建消息接收者 MemberRegisterReceiver。

这里使用@RabbitListener 监听队列名称为 rabbit_mq_member_register_queue_name 的队列，该名称需要对应 Constants.MEMBER_REGISTER_QUEUE_NAME，否则消息将不会成功接收。同时使用@RabbitHandler 注解进行消息的处理，注入了 UserRepository 进行保存会员操作。

```java
@Component
@RabbitListener(queues = "rabbit_mq_member_register_queue_name")
public class MemberRegisterReceiver {
    @Autowired
    private UserRepository userRepository;

    @RabbitHandler
    @Transactional
    public void handler(User member) throws Exception {
        System.out.printf("会员用户名：%s，注册成功，准备创建会员信息...\n", member.getUsername());
        //保存会员消息
        userRepository.save(member);
```

    }
}
```

(8) 新建消息接收者 MemberSendMailReceiver。

该类的创建过程与创建消息接收者 MemberRegisterReceiver 一样，这里需要注意监听的队列名称需要与创建队列时的名称一致。

```
@Component
@RabbitListener(queues = "rabbit_mq_member_send_mail_queue_name")
public class MemberSendMailReceiver {

    @Transactional
    @RabbitHandler
    public void sendMail(User member) throws Exception {
        System.out.printf("会员用户名：%s,注册成功,准备发送邮件...\n", member.getUsername());
    }
}
```

(9) 创建 UserService 接口。

在 service 包下创建对应的 UserService 接口，该接口主要完成会员注册消息的发送。

```
@Service
public class UserServiceImpl implements UserService {
    @Autowired
    private MemberRegisterSender memberRegisterSender;

    public String memberRegister(User member) throws Exception {
        //会员注册
        memberRegisterSender.sendMessage(member);
        return member.getId();
    }
}
```

(10) 创建 UserController 接口。

```
@RestController
public class UserController {

    @Autowired
    private UserService userService;

    @RequestMapping("/registerMember")
    public void registerMember() throws Exception {
        User member = new User();
        member.setId("weixiao123");
        member.setUsername("weixiao");
        member.setPassword("123456");
        member.setEmail("12345678@qq.com");
        userService.memberRegister(member);
    }
}
```

测试运行结果如图 6-31 所示，RabbitMQ 已经启动成功，交换机与两个队列也成功绑定。

访问 http://localhost:8080/registerMember，查看注册信息。因为在两个队列分别进
```

行了保存数据库操作和发送邮件操作,所以消息接收者已经成功接收到消息发送者的消息,并且进行了相应的处理,如图 6-32 所示。

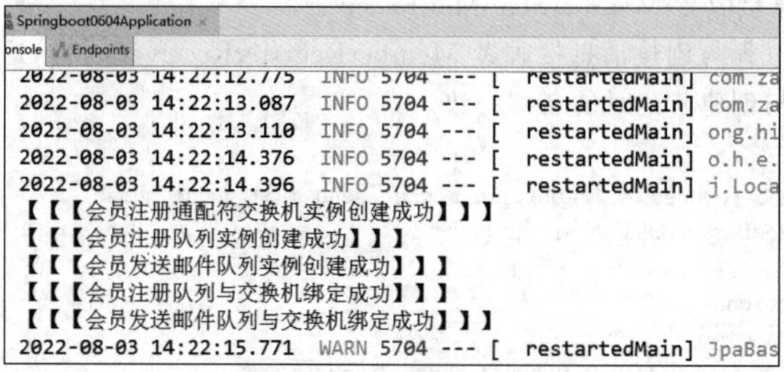

图 6-31 会员注册成功信息

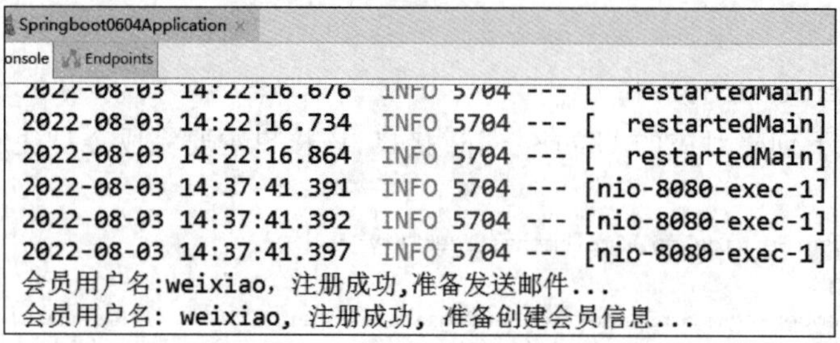

图 6-32 消息发送成功信息

本章小结

本章主要针对 Spring Boot 与 RabbitMQ 消息中间件的整合进行了讲解,包括 RabbitMQ 消息中间件的基本概念与用法、Spring Boot 对 RabbitMQ 工作模式的整合使用。希望通过本章学习,读者能够掌握 Spring Boot 与 RabbitMQ 整合实现消息服务。

在线测试

习题

一、单选题

1. 以下关于一些常用消息中间件的说法,错误的是(   )。
   A. ActiveMQ 是 Apache 出品的基于 AMQP 协议实现的系统
   B. RabbitMQ 是使用 Erlang 语言开发的开源消息队列系统
   C. Kafka 是一种高吞吐量的分布式发布订阅消息系统,它采用 Scala 和 Java 语言编写
   D. RocketMQ 是阿里开源产品,目前是 Apache 的顶级项目,它使用纯 Java 开发,具有高吞吐量、高可用、适合大规模分布式系统应用的特点

2. 以下关于 RabbitMQ 安装过程的说法,正确的是(   )。
   A. 在 Windows 环境下安装 RabbitMQ 消息中间件必须使用 64 位的 Erlang 语言包支持

B. 必须以管理员身份进行 Erlang 语言包安装
C. 在 Windows 环境下首次执行 RabbitMQ 的安装,需要进行 RabbitMQ 系统环境变量设置
D. RabbitMQ 默认提供了两个端口号 5672 和 15672,其中 5672 用作可视化管理端口号,15672 用作服务端口号

3. 以下关于 RabbitMQ 支持的工作模式原理的说法,错误的是(   )。
    A. 在 Work queues 工作模式中,无须交换机即可使用唯一的消息队列进行消息传递
    B. 在 Publish/Subscribe 工作模式中,必须先配置一个 fanout 类型的交换器,不需要指定路由键
    C. 在 Routing 工作模式中,必须先配置一个 direct 类型的交换机,并指定不同的路由键
    D. Headers 工作模式在使用时,必须设置一个 headers 类型的交换机,而不需要设置路由键

4. 在使用 RabbitTemplate 类的 convertAndSend() 方法发送消息时,路由键为 infor.email 会匹配到消息队列(   )。
    A. info.#.email.#   B. info.*.email.*
    C. info.#.email.*   D. info.*.email.#

5. 以下关于 RabbitMQ 实现消息发送接收过程的说法,正确的是(   )。
    A. 在进行消息发送的过程中,实体类消息必须实现 Serializable 接口
    B. 在进行消息发送的过程中,支持 Spring[]、Spring 等类型消息发送
    C. 在 @RabbitListener 注解监听队列消息后,每发送一个消息就会立即被处理
    D. 在使用基于配置类的方式定制消息组件实现消息服务时,可以直接在项目中发送消息,系统会自动创建消息组件

二、多选题

1. 在 RabbitMQ 可视化管理页面中显示的管理面板页面包括(   )。
    A. Channels   B. Exchanges   C. Queues   D. Connections

2. RabbitMQ 内部 SimpleMessageConverter 转换器支持的消息类型包括(   )。
    A. Spring 消息   B. String[] 消息
    C. Serializable 序列化后消息   D. byte[] 消息

3. 以下关于 API 管理类 AmqpAdmin 定制消息发送组件的说法,正确的是(   )。
    A. declareExchange() 方法可以用来声明不同类型的交换器
    B. declareQueue() 方法可以用来声明不同的消息队列
    C. declareRoutingKey() 方法可以用来声明不同的消息路由键
    D. declareBinding() 方法可以用来将消息队列与交换器进行绑定

4. 消息服务在实际开发中的主要用途包括(   )。
    A. 异步处理   B. 应用解耦
    C. 流量削峰   D. 分布式事务管理

5. 以下属于 RabbitMQ 支持的工作模式包括(　　)。
   A. Headers 工作模式　　　　　　　B. Routing 工作模式
   C. Topics 工作模式　　　　　　　　D. RPC 工作模式

三、判断题(对的打"√",错的打"×")

1. 在 RabbitMQ 可视化管理页面的 Exchanges 面板中默认存在 7 个自带的不同类型交换器。(　　)

2. 在 Spring Boot 整合 RabbitMQ 的配置文件中,配置 RabbitMQ 虚拟主机路径/默认可以省略。(　　)

3. 在 Work queues 工作模式中,唯一的消息可以有多个消息消费者,并会被这多个消费者共同消费。(　　)

4. 在 RabbitMQ 消息中间件安装过程中,必须以管理员身份进行 Erlang 语言包安装。(　　)

5. 在 Spring Boot 整合 RabbitMQ 实现消息服务时,可以不必在配置文件中添加任何配置。(　　)

四、填空题

1. 使用 RabbitTemplate 模板类的_____方法手动消费指定队列中的消息。

2. 在多数应用尤其是_____中,消息服务是不可或缺的重要部分。

3. RabbitMQ 安装过程中默认提供了用户名为_____、密码为_____的用户进行登录。

4. 在 Windows 环境下安装 RabbitMQ 消息中间件还需要_____语言包支持。

5. 通过 AmqpAdmin 的_____方法可以定义对应类型的 RabbitMQ 交换器。

五、简答题

简述消息中间件的概念,以及目前常用的一些消息中间件。

第7章

基于Spring Boot+Shiro+Vue开发的前后端分离学生信息管理项目整合实战——后端开发

本章学习目标
- 熟悉 Spring Boot+Shiro 框架整合思路。
- 掌握 Spring Boot+Shiro 框架整合环境构建。
- 掌握 Spring Boot+Shiro 框架整合实现与测试。

通过前面的学习，读者掌握了 Spring Boot、MyBatis 等重要知识点的使用方法。在实际开发中，通常将 Spring Boot 作为开发环境，整合 Shiro、JWT 和 MyBatis 开发后端。本章将对 Spring Boot+Shiro+Vue 前后端分离框架的后端开发进行详细的讲解。

7.1 开发思路整合

本章开发任务要求：
了解基于 Spring Boot+Shiro+Vue 开发的前后端分离学生信息管理项目整合的基本知识和操作。
本章主要内容：
（1）系统设计；
（2）数据库设计；
（3）后端系统设计与实现。
本章系统后端使用 Spring Boot 作为开发环境，搭建的后端项目为 Spring Initializr 项目，整合 Shiro、JWT、MyBatis 和 Redis 开发后端实现各个模块，集成开发环境使用 IntelliJ IDEA 2021.3.2，数据库采用 MySQL。

7.2 系统设计

本章开发任务要求对学生信息管理平台进行总体设计，该平台根据用户角色划分为系统管理员子系统、学生个人信息子系统、辅导员管理子系统、系主任管理子系统、学生处管理子系统和教务处管理子系统等 6 个子系统，旨在通过网络信息化手段办理学生学籍信息管理和学籍异动管理。下面分别说明这 6 个子系统的功能需求与模块划分。

7.2.1 系统功能需求分析

针对高校学生管理信息化、网络化的愿景，最大程度地实现学生信息管理的无纸化办公，尝试性地进行了该学生信息管理平台的整合搭建，旨在提升高校学生信息化管理水平。

（1）系统管理员子系统。

系统管理员子系统要求登录成功后能对用户信息、学生信息进行管理，包括添加用户、删除用户、修改用户，查询学生信息、下载学生信息，查询异动信息、个人用户管理等操作。

（2）学生个人信息子系统。

成功登录后的学生用户具有查看本人学籍信息、床位信息、入学照片信息，个人用户管理、电子签名上传等权限。

（3）辅导员管理子系统。

辅导员管理子系统要求登录成功后能对所管理的班级学生信息进行管理，包括查询学生信息、对学生进行学籍异动操作、提交学籍异动审批、查看学籍异动审批进度、生成学籍异动电子审批单以及个人用户管理、电子签名上传等操作。

辅导员子系统负责学生的学籍异动网上审批提交、线下组织学生异动材料，最后生成由学生本人、辅导员、系主任、学生处处长和教务处处长电子签名的异动审批单。

（4）系主任管理子系统。

成功登录后的用户具有对待审批的学籍异动申请进行网上审批以及个人用户管理、电子签名上传等权限。

（5）学生处管理子系统。

成功登录后的用户具有对待审批的学籍异动申请进行网上审批以及个人用户管理、电子签名上传等权限。

（6）教务处管理子系统。

成功登录后的用户具有对待审批的学籍异动申请进行网上审批以及个人用户管理、电子签名上传等权限。

一般高校学籍异动的休学审批经辅导员审核学生材料提交审批，再经系主任、学生处处长和教务处处长审批后，审批流程即可完成。学籍管理部门根据学生异动审批单即可对学生进行下一步异动操作，包括更改学籍状态、是否离校以及是否在籍等，完成全部学籍异动流程。退学审批手续除上述流程外，还需主管院长审批。

7.2.2 系统模块划分

（1）系统管理员子系统。

系统管理员登录后进入后台管理主页面，可以对用户、学生信息、班级信息、异动信息进行管理。系统管理员子系统的功能模块划分如图 7-1 所示。

（2）学生个人信息子系统。

学生登录后进入学生个人信息子系统，登录成功的学生用户可以查看个人注册信息、入学照片、床位信息以及用户管理，包括上传电子签名和修改密码等。学生个人信息子系统的

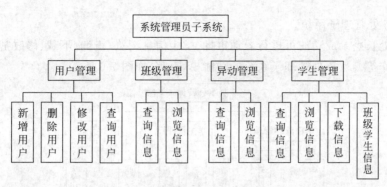

图 7-1 系统管理员子系统功能模块图

功能模块划分如图 7-2 所示。

(3) 辅导员管理子系统。

辅导员登录成功后能对所管理的班级学生信息进行管理，包括查询学生信息、对学生进行学籍异动操作、提交学籍异动审批、查看学籍异动审批进度、生成学籍异动电子审批单以及个人用户管理、上传电子签名等操作。辅导员管理子系统的功能模块划分如图 7-3 所示。

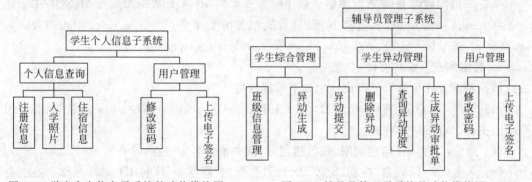

图 7-2 学生个人信息子系统的功能模块图　　图 7-3 辅导员管理子系统的功能模块图

(4) 系主任管理子系统。

系主任登录成功后可进行学籍异动审批、修改密码和上传电子签名等操作等。系主任管理子系统的功能模块划分如图 7-4 所示。

(5) 学生处管理子系统。

学生处登录成功后可进行异动审批、学生信息浏览/查询/下载、修改密码和上传电子签名等操作等。学生处管理子系统的功能模块划分如图 7-5 所示。

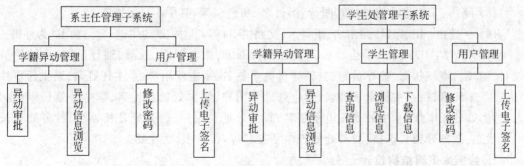

图 7-4 系主任管理子系统的功能模块图　　图 7-5 学生处管理子系统的功能模块图

(6) 教务处管理子系统。

教务处处长登录成功的可进行异动审批、学生信息浏览/查询/下载、修改密码和上传电子签名等操作等。教务处管理子系统的功能模块划分如图 7-6 所示。

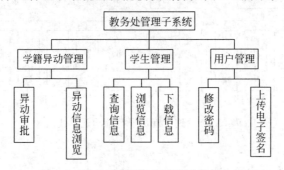

图 7-6　教务处管理子系统的功能模块图

7.2.3　数据库设计

系统采用加载纯 Java 数据库驱动程序的方式连接 MySQL 数据库。在 MySQL 中创建数据库 bunk，并在 bunk 中创建 10 张与系统相关的数据表，即 user、role、student、class、student_photo、class_user、organization、routes、children、suspension 等。

1. 数据库概念结构设计

根据系统设计与分析可以设计出如下数据结构。

（1）用户：用户名、密码、角色 ID、用户权限、是否封号、用户备注、电子签名等，其中用户名为主键，角色 ID 为外键。

（2）角色：角色名、角色权限、角色描述等，其中角色名为主键。

（3）学生：学号、姓名、性别、班级 ID、班级名称、学籍状态、床位 ID、系部名称、专业、是否离校、是否有学籍、年级等，其中学号为主键。

（4）班级：班级 ID、班级名称、年级、是否毕业、系部 ID、系部名称、专业 ID、专业名称等，其中班级 ID 为主键。

（5）学生图像：学号、学生图像等，其中学号为主键。

（6）用户班级：ID、班级 ID、用户名等，其中 ID 为主键，班级 ID、用户名为外键。

（7）系部组织：辅导员用户、系主任用户、系部名称等，其中辅导员用户、系主任用户为外键。

（8）路由：ID、路径、路由名、图标、组件名、角色名等，其中 ID 为主键。

（9）子路由：ID、路径、路由名、组件名、父路由 ID 等，其中 ID 为主键，父路由 ID 为外键。

（10）休学：ID、学号、学生姓名、学生性别、班级、休学理由、系部、辅导员用户、辅导员姓名、辅导员审核状态、辅导员提交、辅导员电子签名、系主任用户、系主任姓名、系主任审核状态、系主任审核、系主任电子签名、学生处处长用户、学生处处长姓名、学生处处长审核状态、学生处处长审核、学生处处长电子签名、教务处处长用户、教务处处长姓名、教务处处长审核状态、教务处处长审核、教务处处长电子签名等，其中 ID 为主键。

2. 数据库逻辑结构设计

将数据库概念结构图转换为 MySQL 数据库所支持的实际数据模型，即数据库的逻辑结构。

用户信息表(user)的设计如表 7-1 所示。

表 7-1　用户信息表

字　　段	含　　义	类　　型	是否为空	说　　明
username	用户名	varchar(20)	否	主键
password	密码	varchar(50)	否	
role	用户角色	varchar(255)	否	外键
permission	权限	varchar(255)		
ban	是否封号	int(11)		
note	用户备注	varchar(50)		用户姓名
autograph	电子签名	mediumtext		

角色信息表(role)的设计如表 7-2 所示。

表 7-2　角色信息表

字　　段	含　　义	类　　型	是否为空	说　　明
role	角色名	varchar(50)	否	主键
permission	角色权限	varchar(11)	否	
descr	角色描述	varchar(50)		

学生信息表(student)的设计如表 7-3 所示。

表 7-3　学生信息表

字　　段	含　　义	类　　型	是否为空	说　　明
student_id	学号	char(14)	否	主键
student_name	姓名	varchar(30)		
sex	性别	char(2)		
class_id	班级 ID	int(11)		
student_state	学籍状态	char(16)		
bunk_id	床位 ID	char(6)		
class_name	班级名称	varchar(50)		
department	系部名称	varchar(50)		
major	专业	varchar(50)		
sflx	是否离校	bit(1)		
sfyxj	是否有学籍	bit(1)		
grade	年级	varchar(4)		

班级信息表(class)的设计如表 7-4 所示。

表 7-4　班级信息表

字　　段	含　　义	类　　型	是否为空	说　　明
class_id	班级 ID	int(11)	否	主键
class_name	班级名称	varchar(18)	否	
grade	当前所在年级	varchar(18)	否	
is_graduated	是否毕业	bit(1)	否	
department_code	系部 ID	int(11)	否	
department	系部名称	varchar(255)	否	
majoy_code	专业 ID	char(6)	否	
majoy	专业名称	varchar(50)	否	

学生图像信息表（student_photo）的设计如表 7-5 所示。

表 7-5　学生图像信息表

字　段	含　义	类　型	是否为空	说　明
student_id	学号	char(14)	否	主键
student_photo	学生图像	mediumtext		

用户班级信息表（class_user）的设计如表 7-6 所示。

表 7-6　用户班级信息表

字　段	含　义	类　型	是否为空	说　明
id	ID	int(11)	否	主键
class_id	班级 ID	int(12)	否	外键
username	用户名	varchar(12)	否	外键

系部组织信息表（organization）的设计如表 7-7 所示。

表 7-7　系部组织信息表

字　段	含　义	类　型	是否为空	说　明
counsellor	辅导员用户	varchar(50)	否	外键
depprincipal	系主任用户	varchar(50)	否	外键
department	系部名称	varchar(50)	否	

路由信息表（routes）的设计如表 7-8 所示。

表 7-8　路由信息表

字　段	含　义	类　型	是否为空	说　明
pathroleId	路径角色 ID	int(11)	否	主键
path	路径	varchar(255)	否	
name	路由名	varchar(255)	否	
icon	图标	varchar(255)	否	
component	组件名	varchar(255)	否	
role	角色	varchar(255)	否	外键

子路由信息表（children）的设计如表 7-9 所示。

表 7-9　子路由信息表

字　段	含　义	类　型	是否为空	说　明
id	ID	int(11)	否	主键
cpath	路径	varchar(255)	否	
cname	路由名	varchar(255)	否	
ccomponent	组件名	varchar(255)	否	
parentId	父路由 ID	int(11)	否	外键

休学信息表（suspension）的设计如表 7-10 所示。

表 7-10　休学信息表

字　段	含　义	类　型	是否为空	说　明
id	ID	int(11)	否	主键
studentId	学号	varchar(12)		

续表

字段	含义	类型	是否为空	说明
studentName	学生姓名	varchar(50)		
sex	学生性别	varchar(2)		
className	班级	int(11)		
reason	休学理由	varchar(255)		
department	系部	varchar(50)		
counsellor	辅导员用户名	varchar(50)		
cusnote	辅导员姓名	varchar(50)		
cstate	辅导员提交状态	varchar(10)		
csubmit	辅导员提交	bit		
cautograph	辅导员电子签名	mediumtext		
depprincipal	系主任用户名	varchar(50)		
depnote	系主任姓名	varchar(50)		
dstate	系主任审批状态	varchar(10)		
dsubmit	系主任审批	bit		
dautograph	系主任电子签名	mediumtext		
stumanager	学生处长用户名	varchar(50)		
stumnote	学生处长姓名	varchar(50)		
stumstate	学生处处长审批状态	varchar(10)		
stumsubmit	学生处处长审批	bit		
stumautograph	学生处处长电子签名	mediumtext		
dean	教务处长用户名	varchar(50)		
deannote	教务处处长姓名	varchar(50)		
deanstate	教务处处长审批状态	varchar(10)		
deansubmit	教务处处长审批	bit		
deanautograph	教务处处长电子签名	mediumtext		
date	休学时间	datetime		

3. 创建数据表

根据以上所述的逻辑结构创建数据表,具体数据库脚本如下所示。

```
Bunk.sql 数据库脚本
-- --------------------
-- Table structure for children
-- --------------------
DROP TABLE IF EXISTS `children`;
CREATE TABLE `children` (
  `parentId` int(11) DEFAULT NULL,
  `cpath` varchar(255) DEFAULT NULL,
  `cname` varchar(255) DEFAULT NULL,
  `ccomponent` varchar(255) DEFAULT NULL,
  `id` int(11) NOT NULL AUTO_INCREMENT,
  PRIMARY KEY (`id`),
  KEY `children` (`parentId`),
  CONSTRAINT `c_r` FOREIGN KEY (`parentId`) REFERENCES `routes` (`pathroleId`)
) ENGINE = InnoDB AUTO_INCREMENT = 28 DEFAULT CHARSET = utf8;
```

```sql
-- ----------------------------
-- Table structure for class
-- ----------------------------
DROP TABLE IF EXISTS `class`;
CREATE TABLE `class` (
  `class_id` int(11) NOT NULL,
  `class_name` varchar(18) DEFAULT NULL,
  `grade` varchar(18) DEFAULT NULL,
  `is_graduated` bit(1) DEFAULT NULL,
  `department_code` int(11) DEFAULT NULL,
  `majoy_code` char(6) DEFAULT NULL,
  `majoy` varchar(50) DEFAULT NULL,
  `dapartment` varchar(255) DEFAULT NULL,
  PRIMARY KEY (`class_id`),
  KEY `dapartment` (`dapartment`)
) ENGINE = InnoDB DEFAULT CHARSET = utf8;
-- ----------------------------
-- Table structure for class_user
-- ----------------------------
DROP TABLE IF EXISTS `class_user`;
CREATE TABLE `class_user` (
  `id` int(11) NOT NULL AUTO_INCREMENT,
  `class_id` int(12) NOT NULL,
  `username` varchar(12) NOT NULL,
  PRIMARY KEY (`id`),
  KEY `c` (`class_id`),
  KEY `u` (`username`),
  CONSTRAINT `c` FOREIGN KEY (`class_id`) REFERENCES `class` (`class_id`),
  CONSTRAINT `u` FOREIGN KEY (`username`) REFERENCES `user` (`username`)
) ENGINE = InnoDB AUTO_INCREMENT = 10 DEFAULT CHARSET = utf8;
-- ----------------------------
-- Table structure for organization
-- ----------------------------
DROP TABLE IF EXISTS `organization`;
CREATE TABLE `organization` (
  `counselor` varchar(50) NOT NULL,
  `depprincipal` varchar(50) NOT NULL,
  `department` varchar(50) NOT NULL,
  KEY `c_u` (`counselor`),
  KEY `d_u` (`depprincipal`),
  KEY `d_c` (`department`),
  CONSTRAINT `c_u` FOREIGN KEY (`counselor`) REFERENCES `user` (`username`),
  CONSTRAINT `d_c` FOREIGN KEY (`department`) REFERENCES `department` (`department`),
  CONSTRAINT `d_u` FOREIGN KEY (`depprincipal`) REFERENCES `user` (`username`)
) ENGINE = InnoDB DEFAULT CHARSET = utf8;
-- ----------------------------
-- Table structure for role
-- ----------------------------
DROP TABLE IF EXISTS `role`;
CREATE TABLE `role` (
  `role` varchar(50) NOT NULL,
```

```sql
  `permission` varchar(11) DEFAULT NULL,
  `descr` varchar(50) DEFAULT NULL,
  PRIMARY KEY (`role`),
  KEY `role_right` (`permission`)
) ENGINE = InnoDB DEFAULT CHARSET = utf8;
-- ----------------------
-- Table structure for routes
-- ----------------------
DROP TABLE IF EXISTS `route`;
CREATE TABLE `route` (
  `path` varchar(255) NOT NULL,
  `name` varchar(255) DEFAULT NULL,
  `icon` varchar(255) DEFAULT NULL,
  `component` varchar(255) DEFAULT NULL,
  `children` varchar(255) DEFAULT NULL,
  `role` varchar(255) DEFAULT NULL,
  `pathroleId` int(11) NOT NULL AUTO_INCREMENT,
  PRIMARY KEY (`pathroleId`),
  KEY `role` (`role`),
  KEY `c` (`children`),
  CONSTRAINT `role` FOREIGN KEY (`role`) REFERENCES `role` (`role`)
) ENGINE = InnoDB AUTO_INCREMENT = 21 DEFAULT CHARSET = utf8;
-- ----------------------
-- Table structure for student
-- ----------------------
DROP TABLE IF EXISTS `student`;
CREATE TABLE `student` (
  `student_id` char(14) NOT NULL,
  `student_name` varchar(30) DEFAULT NULL,
  `sex` char(2) DEFAULT NULL,
  `class_id` int(11) DEFAULT NULL,
  `student_state` char(16) DEFAULT NULL,
  `bunk_id` varchar(255) DEFAULT NULL,
  `class_name` varchar(255) DEFAULT NULL,
  `department` varchar(255) DEFAULT NULL,
  `major` varchar(255) DEFAULT NULL,
  `sflx` bit(1) DEFAULT NULL,
  `sfyxj` bit(1) DEFAULT NULL,
  `grade` varchar(4) DEFAULT NULL,
  PRIMARY KEY (`student_id`),
  KEY `FK_Reference_2` (`class_id`) USING BTREE
) ENGINE = InnoDB DEFAULT CHARSET = utf8;
-- ----------------------
-- Table structure for student_photo
-- ----------------------
DROP TABLE IF EXISTS `student_photo`;
CREATE TABLE `student_photo` (
  `student_id` char(14) NOT NULL,
  `student_photo` mediumtext,
  PRIMARY KEY (`student_id`)
) ENGINE = InnoDB DEFAULT CHARSET = utf8;
```

```sql
-- ----------------------------
-- Table structure for suspension
-- ----------------------------
DROP TABLE IF EXISTS `suspension`;
CREATE TABLE `suspension` (
  `id` int(11) NOT NULL AUTO_INCREMENT,
  `studentId` varchar(12) DEFAULT NULL,
  `studentName` varchar(50) DEFAULT NULL,
  `sex` varchar(2) DEFAULT NULL,
  `className` varchar(50) DEFAULT NULL,
  `reason` varchar(255) DEFAULT NULL,
  `department` varchar(50) DEFAULT NULL,
  `counselor` varchar(50) DEFAULT NULL,
  `cusnote` varchar(50) DEFAULT NULL,
  `cstate` varchar(10) DEFAULT NULL,
  `csubmit` bit(1) DEFAULT NULL,
  `cautograph` mediumtext,
  `depprincipal` varchar(50) DEFAULT NULL,
  `depnote` varchar(50) DEFAULT NULL,
  `dstate` varchar(10) DEFAULT NULL,
  `dsubmit` bit(1) DEFAULT NULL,
  `dautograph` mediumtext,
  `stumanager` varchar(50) DEFAULT NULL,
  `stumnote` varchar(50) DEFAULT NULL,
  `stumstate` varchar(10) DEFAULT NULL,
  `stumsubmit` bit(1) DEFAULT NULL,
  `stumautograph` mediumtext,
  `dean` varchar(50) DEFAULT NULL,
  `deannote` varchar(50) DEFAULT NULL,
  `deanstate` varchar(10) DEFAULT NULL,
  `deansubmit` bit(1) DEFAULT NULL,
  `deanautograph` mediumtext,
  `date` datetime DEFAULT NULL,
  PRIMARY KEY (`id`)
) ENGINE = InnoDB AUTO_INCREMENT = 59 DEFAULT CHARSET = utf8;
-- ----------------------------
-- Table structure for user
-- ----------------------------
DROP TABLE IF EXISTS `user`;
CREATE TABLE `user` (
  `username` varchar(20) NOT NULL,
  `password` varchar(50) NOT NULL,
  `role` varchar(255) DEFAULT NULL,
  `permission` varchar(255) DEFAULT NULL,
  `ban` int(11) DEFAULT NULL,
  `note` varchar(50) DEFAULT NULL,
  `autograph` mediumtext,
  PRIMARY KEY (`username`),
  KEY `u_r` (`role`),
  CONSTRAINT `u_r` FOREIGN KEY (`role`) REFERENCES `role` (`role`)
) ENGINE = InnoDB DEFAULT CHARSET = utf8;
```

4. 数据库表完整性约束设计

根据数据表的主外键关联，生成数据库表约束模型，以保证数据的正确性和完整性，数据模型如图 7-7 所示。

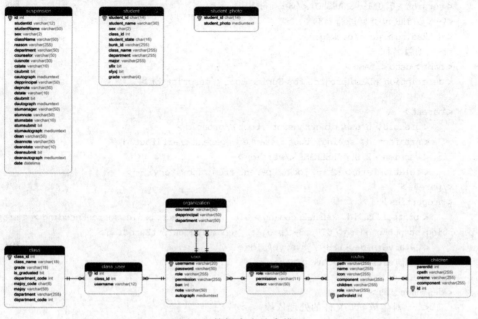

图 7-7　数据库表约束模型

7.3　后端系统环境搭建

搭建 Spring Initializr 项目，配置项目环境，使用 Maven 的 pom.xml 文件对项目所依赖的 JAR 包进行导入，对项目后端模块进行设计，并对项目相关文件进行配置。

7.3.1　使用 Maven 组件为项目添加依赖 JAR 包

新建一个 Spring Initializr 项目，将其命名为 Demo，配置项目坐标 group 和 artifact，选择开发语言 Java，设置项目路径、包名和打包类型（选 JAR 类型）。完成配置后即可在 Demo 项目中开发系统，系统采用纯 Java 数据库驱动程序连接 MySQL 数据库。项目的 Maven 的依赖配置文件 pom.xml 是本项目中需要使用的所有 JAR 包文件名称和版本的配置，通过加载后会自动保存到本地仓库以备项目开发。系统所使用的核心功能依赖包括 c3p0 数据库连接池依赖、MySQL 数据库驱动依赖、MyBatis 框架依赖、Redis 缓存依赖、Shiro 登录授权验证依赖、Web 依赖以及生成 PDF 文件依赖等。

pom.xml 文件代码

```
<?xml version = "1.0" encoding = "UTF - 8"?>
< project xmlns = "http://maven.apache.org/POM/4.0.0"
         xmlns:xsi = "http://www.w3.org/2001/XMLSchema - instance"
         xsi:schemaLocation = "http://maven.apache.org/POM/4.0.0 http://maven.apache.org/
xsd/maven - 4.0.0.xsd">
```

```xml
<modelVersion>4.0.0</modelVersion>
<groupId>com.imooc</groupId>
<artifactId>demo</artifactId>
<version>0.0.1-SNAPSHOT</version>
<!-- 以JAR包类型进行打包 -->
<packaging>jar</packaging>
<!-- 项目名称 -->
<name>demo</name>
<description>Demo project for Spring Boot</description>

<parent>
    <groupId>org.springframework.boot</groupId>
    <artifactId>spring-boot-starter-parent</artifactId>
    <version>2.0.4.RELEASE</version>
    <relativePath/> <!-- lookup parent from repository -->
</parent>
<properties>
    <project.build.sourceEncoding>UTF-8</project.build.sourceEncoding>
    <project.reporting.outputEncoding>UTF-8</project.reporting.outputEncoding>
    <java.version>1.8</java.version>
    <skipTests>true</skipTests>
</properties>
<dependencies>
    <!-- 对生成PDF文件的处理依赖 -->
    <dependency>
        <groupId>com.itextpdf</groupId>
        <artifactId>itextpdf</artifactId>
        <version>5.5.13.2</version>
    </dependency>
    <dependency>
        <groupId>com.itextpdf</groupId>
        <artifactId>itext-asian</artifactId>
        <version>5.2.0</version>
    </dependency>
    <!-- @Slf4j -->
    <dependency>
        <groupId>org.projectlombok</groupId>
        <artifactId>lombok</artifactId>
        <version>1.16.18</version>
        <scope>provided</scope>
    </dependency>
    <dependency>
        <groupId>com.itextpdf.tool</groupId>
        <artifactId>xmlworker</artifactId>
        <version>5.5.13.1</version>
    </dependency>
    <dependency>
        <groupId>org.springframework.boot</groupId>
        <artifactId>spring-boot-starter-web</artifactId>
    </dependency>
    <dependency>
        <groupId>org.mybatis.spring.boot</groupId>
        <artifactId>mybatis-spring-boot-starter</artifactId>
```

```xml
        <version>1.3.2</version>
</dependency>
<dependency>
        <groupId>mysql</groupId>
        <artifactId>mysql-connector-java</artifactId>
        <version>5.1.46</version>
</dependency>
<dependency>
        <groupId>com.mchange</groupId>
        <artifactId>c3p0</artifactId>
        <version>0.9.5.2</version>
</dependency>
<dependency>
        <groupId>org.springframework.boot</groupId>
        <artifactId>spring-boot-starter-test</artifactId>
        <scope>test</scope>
</dependency>
<dependency>
        <groupId>com.alibaba</groupId>
        <artifactId>fastjson</artifactId>
        <version>1.2.8</version>
</dependency>
<dependency>
        <groupId>com.alibaba</groupId>
        <artifactId>druid</artifactId>
        <version>1.0.27</version>
</dependency>
<!-- 登录验证框架的依赖 -->
<dependency>
        <groupId>org.apache.shiro</groupId>
        <artifactId>shiro-spring</artifactId>
        <version>1.3.2</version>
</dependency>
<dependency>
        <groupId>com.auth0</groupId>
        <artifactId>java-jwt</artifactId>
        <version>3.2.0</version>
</dependency>
<!-- redis -->
<dependency>
        <groupId>org.springframework.data</groupId>
        <artifactId>spring-data-redis</artifactId>
</dependency>
<!-- shiro-redis -->
<dependency>
        <groupId>org.crazycake</groupId>
        <artifactId>shiro-redis</artifactId>
        <version>2.4.2.1-RELEASE</version>
</dependency>
<dependency>
        <groupId>org.apache.shiro</groupId>
        <artifactId>shiro-core</artifactId>
        <version>1.3.2</version>
</dependency>
```

```xml
        <dependency>
            <groupId>org.apache.tomcat.embed</groupId>
            <artifactId>tomcat-embed-core</artifactId>
            <version>9.0.64</version>
        </dependency>
        <dependency>
            <groupId>org.apache.shiro</groupId>
            <artifactId>shiro-web</artifactId>
            <version>1.3.2</version>
        </dependency>
    </dependencies>
    <build>
        <plugins>
            <plugin>
                <groupId>org.springframework.boot</groupId>
                <artifactId>spring-boot-maven-plugin</artifactId>
            </plugin>
        </plugins>
    </build>
</project>
```

7.3.2 项目的目录结构

Demo项目的目录结构如图7-8所示。

后端业务采用经典的三层架构思想,即将项目划分为控制层Controller、业务逻辑层Service、数据访问层DAO进行开发,同时项目引入登录授权Shiro框架、Redis缓存和自定义的配置类及工具类等。

(1) web包。

系统的控制器类都在该包中,即网络资源,对用户角色的权限控制主要是对控制器类的访问控制。

(2) config包。

config包中存放了DataSourceConfiguration、SessionFactoryConfiguration和TranscationManagementConfiguration三个Java配置类。其中,DataSourceConfiguration的作用是对数据源进行配置;SessionFactoryConfiguration的作用是生成SqlSessionFactoryBean对象sqlSessionFactory,即生成Session工厂;TranscationManagementConfiguration的作用是开启事务自动装配。

(3) dao包。

dao包中的mapper文件用于实现数据库的接口,各接口的方法名与对应MyBatis的XML文

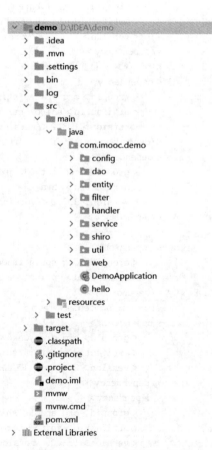

图7-8 Demo项目的目录结构

件 id 一致。

(4) entity 包。

持久化类存放在此包中。

(5) service 包。

service 包存放后台相关业务层的接口，包括 serviceImpl 类、实现 Service 接口的方法和调用 DAO 接口的方法。

(6) util 包。

util 包中存放系统的工具类。

(7) shiro 包。

shiro 包中创建了 CustomRealm 类，该类的作用是查找相关数据源，作为安全管理间的桥梁，对请求数据进行认证、授权等操作。另外存放了 ShiroConfig 配置类，该类的主要作用是创建安全管理器、配置 Shiro 过滤器工厂和开启对 Shiro 注解的支持等。

(8) filter 包。

filter 包存放了 JWTFilter 类，对资源请求进行 JWT 过滤，完成登录或拦截操作。

(9) handler 包。

handler 包存放了 GlobalExceptionHandler 类，实现全局异常处理。

(10) DemoApplication 类。

DemoApplication 类为项目启动类。

7.3.3 项目的配置文件

1. application.properties

application.properties 是 Spring Boot 框架中的一个全局配置文件，是项目的核心配置文件。项目创建成功后，会在其资源文件夹 resources 下自动生成一个 application.properties 的空配置文件。在该配置文件中，编写端口号配置、数据库连接、日志记录、MyBatis 配置等代码。

```
application.properties 文件代码
--------------------------------------------------------------------------
server.port = 8081
jdbc.driver = com.mysql.jdbc.Driver
jdbc.url = jdbc:mysql://127.0.0.1:3306/bunk?useUnicode = true&characterEncoding = utf8&useSSL = false
jdbc.username = root
jdbc.password = 123456
spring.datasource.initialSize = 5
spring.datasource.MinPoolSize = 5
spring.datasource.MaxPoolSize = 20
#log
logging.file = log/log.log
logging.level.com.howie.shiro.mapper = DEBUG
logging.level.org.springframework.web = DEBUG
#mybatis
mybatis_config_file = mybatis-config.xml
mapper_path = /mapper/**.xml
entity_package = com.imooc.demo.entity
```

```
#redis
#RedisManager 类默认参数:
#private String host = "127.0.0.1";
#private int port = 6379;
#private int expire = 0;
#private int timeout = 0;
#private String password = "";
#private static JedisPool jedisPool = null;
```

2. mybatis-config.xml

该配置文件用于配置数据库连接属性。由于在 application.properties 中已经配置了数据源，因此在 mybatis-config.xml 中只对实体类别名进行配置。

mybatis-config.xml 文件代码

```xml
<?xml version="1.0" encoding="UTF-8"?>
<!DOCTYPE configuration PUBLIC "-//mybatis.org//DTD Config 3.0//EN" "http://mybatis.org/dtd/mybatis-3-config.dtd">
<configuration>
    <settings>
        <setting name="cacheEnabled" value="true"/>
        <setting name="lazyLoadingEnabled" value="true"/>
        <setting name="multipleResultSetsEnabled" value="true"/>
        <setting name="useColumnLabel" value="true"/>
        <setting name="useGeneratedKeys" value="true"/>
        <setting name="autoMappingBehavior" value="PARTIAL"/>
        <setting name="defaultExecutorType" value="SIMPLE"/>
        <setting name="defaultStatementTimeout" value="25"/>
        <setting name="safeRowBoundsEnabled" value="false"/>
        <setting name="mapUnderscoreToCamelCase" value="true"/>
        <setting name="localCacheScope" value="SESSION"/>
        <setting name="jdbcTypeForNull" value="OTHER"/><setting name="lazyLoadTriggerMethods" value="equals,clone,hashCode,toString"/>
    </settings>
    <!-- 配置别名 -->
    <typeAliases>
        <package name="com.imooc.demo.entity"/>
    </typeAliases>
</configuration>
```

7.3.4 项目的配置类

Spring Boot 项目将@Configuration、@MapperScan 等注解注入 Java 类中，取代 SSM 架构的 XML 配置文件，完成 Spring 对数据库连接以及 DAO 接口文件、MyBatis 的数据库 XML 配置文件的注册装配工作。

1. 数据库连接配置类 DataSourceConfiguration

在该配置类上添加@Configuration 注解和@MapperScan("com.imooc.demo.dao")注解，其中@MapperScan 完成对 com.imooc.demo.dao 包中的接口文件扫描，将接口文件注入 Spring 容器中。在配置类中声明一个返回值类型为 ComboPooledDataSource 的 createDataSource()方法，再通过@Value 注解将 application.properties 文件中的 jdbc 驱动类 jdbc.driver、数据库链接地址 jdbc.url、用户名 jdbc.username、密码 jdbc.password 的值

等注入方法中。ComboPooledDataSource 为 c3p0 数据库连接池配置类（在 pom.xml 文件中已经导入了 c3p0 的依赖）。

c3p0 的依赖
```xml
<dependency>
    <groupId>com.mchange</groupId>
    <artifactId>c3p0</artifactId>
    <version>0.9.5.2</version>
</dependency>
```

DataSourceConfiguration.java 文件代码
```java
package com.imooc.demo.config.dao;
import com.mchange.v2.c3p0.ComboPooledDataSource;
import org.mybatis.spring.annotation.MapperScan;
import org.springframework.beans.factory.annotation.Value;
import org.springframework.context.annotation.Bean;
import org.springframework.context.annotation.Configuration;
import java.beans.PropertyVetoException;
@Configuration
//配置 MyBatis mapper 的扫描路径
@MapperScan("com.imooc.demo.dao")
public class DataSourceConfiguration {
    @Value("${jdbc.driver}")
    private String jdbcDriver;
    @Value("${jdbc.url}")
    private String jdbcUrl;
    @Value("${jdbc.username}")
    private String jdbcUserName;
    @Value("${jdbc.password}")
    private String jdbcPassWord;
    @Value("${spring.datasource.initialSize}")
    private int initialSize;
    @Value("${spring.datasource.MinPoolSize}")
    private int minPoolSize;
    @Value("${spring.datasource.MaxPoolSize}")
    private int maxPoolSize;

    @Bean(name = "dataSource")
public ComboPooledDataSource createDataSource() throws PropertyVetoException {
        ComboPooledDataSource dataSource = new ComboPooledDataSource();
        dataSource.setDriverClass(jdbcDriver.trim());      //数据库驱动
        dataSource.setJdbcUrl(jdbcUrl.trim());             //数据库地址
        dataSource.setUser(jdbcUserName.trim());           //用户名
        dataSource.setPassword(jdbcPassWord.trim());       //密码
        dataSource.setInitialPoolSize(initialSize);        //初始化连接数
        dataSource.setMinPoolSize(minPoolSize);            //最小连接数
        dataSource.setMaxPoolSize(maxPoolSize);            //最大连接数
        dataSource.setAutoCommitOnClose(false);            //关闭连接后不自动 commit
        return dataSource;
    }
}
```

2. Session 工厂配置类 SessionFactoryConfiguration

数据库连接成功后需要配置生产 SqlSession 对象的工厂类 SqlSessionFactoryBean，它负责以工厂模式创建 SqlSession 对象。SqlSession 是 MyBatis 的关键对象，它是应用程序与持久层之间交互操作的一个单线程对象，类似于 JDBC 中的 Connection。

SessionFactoryConfiguration.java 文件代码

```java
package com.imooc.demo.config.dao;
import org.mybatis.spring.SqlSessionFactoryBean;
import org.springframework.beans.factory.annotation.Autowired;
import org.springframework.beans.factory.annotation.Qualifier;
import org.springframework.beans.factory.annotation.Value;
import org.springframework.context.annotation.Bean;
import org.springframework.context.annotation.Configuration;
import org.springframework.core.io.ClassPathResource;
import org.springframework.core.io.support.PathMatchingResourcePatternResolver;
import javax.sql.DataSource;
import java.io.IOException;
@Configuration
public class SessionFactoryConfiguration {
    @Value("${mybatis_config_file}")
    private String mybatisConfigFilePath;
    @Value("${mapper_path}")
    private String mapperPath;
    @Autowired
    @Qualifier("dataSource")
    private DataSource dataSource;
    @Value("${entity_package}")
    private String entityPackage;
    @Bean(name = "sqlSessionFactory")
    public SqlSessionFactoryBean craeteSqlSessionFactoryBean() throws IOException {
        SqlSessionFactoryBean sqlSessionFactoryBean = new SqlSessionFactoryBean();
        sqlSessionFactoryBean.setConfigLocation(new ClassPathResource(mybatisConfigFilePath));
        PathMatchingResourcePatternResolver resolver = new PathMatchingResourcePatternResolver();
        String packageSearchPath = PathMatchingResourcePatternResolver.CLASSPATH_ALL_URL_PREFIX + mapperPath;
        sqlSessionFactoryBean.setMapperLocations(resolver.getResources(packageSearchPath));
        sqlSessionFactoryBean.setDataSource(dataSource);
        sqlSessionFactoryBean.setTypeAliasesPackage(entityPackage);
        return sqlSessionFactoryBean;
    }
}
```

在 application.properties 文件中配置 MyBatis 文件的主要路径：

```
#mybatis
mybatis_config_file=mybatis-config.xml
mapper_path=/mapper/**.xml
entity_package=com.imooc.demo.entity
```

在 SessionFactoryConfiguration 类代码中，通过 @Value 注解将 application.properties 文件中的 mybatis-config.xml 配置文件路径、MyBatis 的 *Mapper.xml 文件所在路径、实体类所在路径以及数据源对象 dataSource 注入 craeteSqlSessionFactoryBean 方法中。该方

法返回一个 SqlSessionFactoryBean 对象，整合了数据源、MyBatis 的数据库 XML 配置文件及实体类等，其中该配置类通过 @Configuration 注解注入 Spring 容器中。

3. Redis 配置类 RedisConfiguration

作为一个中小型的 Web 应用开发，一般的用户级别都会是以千、万为单位的，所以系统的抗并发能力和访问抗压能力是一个关键问题。访问用户多、访问频次高的情况下，每次请求都要和数据库连接，很容易导致系统崩溃。考虑到这些因素，本项目采用 Redis 数据库存储一些高频发的访问数据，以达到数据缓存的目的。众所周知 Redis 是一个高性能的、基于内存的 key-value 数据库，其性能极高，这样会大大减轻系统对 MySQL 数据库的访问压力，也会显著提高系统的整体性能。

Redis 配置类的实现步骤如下：

（1）在 pom.xml 中引入 spring-data-redis 依赖。

（2）在 Redis 配置类 RedisConfiguration 中声明一个 RedisTemplate<String,Object>对象 template，注入 RedisConnectionFactory 对象 redisConnectionFactory，然后对 template 进行序列化配置。通用的 key 和 hash 的 key 都采用 string 的序列化方式，通用的 value 和 hash 的 value 都采用 json 的序列化方式。

（3）方法返回配置好的 RedisTemplate<String,Object>对象 template。

pom.xml 的 redis 依赖文件代码

```
<dependency>
<groupId>org.springframework.data</groupId>
<artifactId>spring-data-redis</artifactId>
</dependency>
```

SessionFactoryConfiguration.java 文件代码

```
package com.imooc.demo.config.dao;
import org.springframework.context.annotation.Bean;
import org.springframework.context.annotation.Configuration;
import org.springframework.data.redis.connection.RedisConnectionFactory;
import org.springframework.data.redis.core.RedisTemplate;
import org.springframework.data.redis.serializer.GenericJackson2JsonRedisSerializer;
import org.springframework.data.redis.serializer.StringRedisSerializer;
@Configuration
public class RedisConfiguration {
    @Bean
    public RedisTemplate<String,Object> redisTemplate(RedisConnectionFactory redisConnectionFactory){
        RedisTemplate<String,Object> template = new RedisTemplate<String,Object>();
        template.setConnectionFactory(redisConnectionFactory);
        //key采用string的序列化方式
        template.setKeySerializer(new StringRedisSerializer());
        //value采用jackson的序列化方式
        template.setValueSerializer(new GenericJackson2JsonRedisSerializer());
        //hash的key采用string的序列化方式
        template.setHashKeySerializer(new StringRedisSerializer());
        //hash的value采用jackson的序列化方式
        template.setHashValueSerializer(new GenericJackson2JsonRedisSerializer());
```

```
        return template;
    }
}
```

7.4　Apache Shiro 的工作机制和配置类设计

本系统运用 Apache Shiro 组件完成前后端分离框架的核心功能。由于不使用 Session 会话机制，因此资源访问请求的身份验证和权限认证就成了前后端分离框架的主要工作，其内容包括用户登录、授权认证、Redis 缓存设计、统一异常处理和工具类设计等。

7.4.1　Shiro 的工作机制

该前后端分离项目的登录授权工作由 Apache Shiro 完成。

Apache Shiro 是一个强大且灵活的开源安全框架，它能够快捷有效地完成身份认证、授权、企业会话管理和加密。相比 Spring Security 框架而言，它更加简单灵活，并且对 spring 依赖较弱。Shiro 可以实现 Web、C/S、分布式等系统权限管理。Shiro 的完整架构如图 7-9 所示。

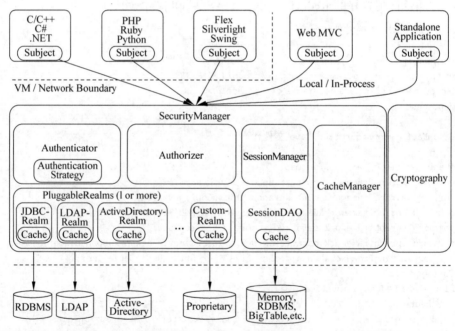

图 7-9　Shiro 的完整架构

1. Shiro 核心组件

Subject：主体，即当前操作用户。

SecurityManager：安全管理器。它是 Shiro 框架的核心，采用典型的 Facade 模式。Shiro 通过 SecurityManager 管理内部组件实例，并通过它来提供安全管理的各种服务。

Realm：领域。它作为 Shiro 与应用安全数据间的"桥梁"或"连接器"。也就是说，当对用户执行认证（登录）和授权（访问控制）验证时，Shiro 会从应用配置的 Realm 中查找用户及其权限信息。

Authenticator：认证器。AuthenticatonStrategy 如果存在多个 Realm，则按具体的策略进行登录控制。例如，如果有一个 Realm 成功即可登录，必须所有 Realm 都成功才能登录等。

Authorizer：授权器。它决定 Subject 能拥有什么样角色或者权限。

SessionManager：Session 管理器。它负责创建和管理用户 Session。通过设置这个管理器，Shiro 可以在任何环境下使用 Session。本框架中设置 Shiro 自带的 Session 为关闭状态，改用 JWT(JSON Web Token)格式的 Token(令牌)作为前后端的联系纽带。

CacheManager：缓存管理器。它可以减少不必要的后台访问，提高应用效率，增加用户体验。

Cryptography：密码体系。Shiro API 可以大幅简化 Java API 中烦琐的密码加密。

2. Shiro 认证流程

Shiro 认证流程如图 7-10 所示，其实现步骤如下。

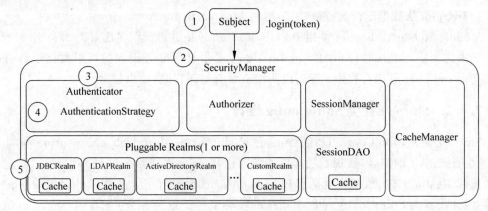

图 7-10 Shiro 认证流程

(1) Subject(主体)请求认证，调用 Subject.login(token)。

(2) SecurityManager(安全管理器)执行认证。

(3) SecurityManager 通过 ModularRealmAuthenticator 进行认证。

(4) ModularRealmAuthenticator 将 Token 传给 Realm，Realm 根据 Token 中的用户信息从数据库查询用户信息(包括身份和凭证)。

Realm 如果查询不到用户则给 ModularRealmAuthenticator 返回 null，ModularRealmAuthenticator 抛出异常(用户不存在)；Realm 如果查询到用户则给 ModularRealmAuthenticator 返回 AuthenticationInfo(认证信息)。

(5) ModularRealmAuthenticator 将 AuthenticationInfo(认证信息)进行凭证(密码)比对。如果比对一致则认证通过，如果比对不一致则抛出异常(凭证错误)。

3. Shiro 授权流程

Shiro 授权流程如图 7-11 所示，其实现步骤如下。

(1) 对 Subject 进行授权，调用方法 isPermitted(" * ")或 hasRole(" * ")。

(2) SecurityManager 执行授权，通过 ModularRealmAuthorizer 执行授权。

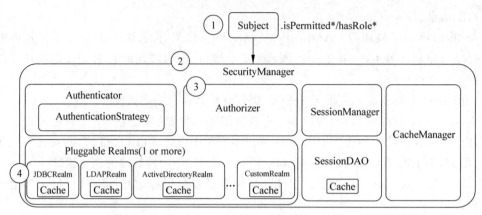

图 7-11 Shiro 授权流程

(3) ModularRealmAuthorizer 执行 Realm(自定义的 CustomRealm)从数据库查询权限数据调用 Realm 的授权方法 doGetAuthorizationInfo。

(4) Realm 从数据库查询权限数据，返回 ModularRealmAuthorizer。
ModularRealmAuthorizer 调用 PermissionResolver 进行权限串比对。

如果比对后，isPermitted 中的"permission 串"在 Realm 查询到的权限数据中，则说明用户访问 permission 串有权限；否则说明用户没有访问权限，抛出异常。

7.4.2 Shiro 配置类 ShiroConfig 设计

Shiro 配置类 ShiroConfig 设计就是通过对 shiro-core、shiro-web、shiro-spring 和 shiro-redis 等依赖 JAR 的整合，实现系统的资源访问拦截、安全管理器 SecurityManager 创建、安全数据源 Realm 创建、数据库会话管理 SessionManager 整合、缓存管理 CacheManager 整合等，因此配置类 ShiroConfig 是后端系统的核心组件。创建 Shiro 配置类 ShiroConfig 一般分为以下 4 个步骤：

（1）创建安全数据源 CustomRealm。
（2）创建安全管理器。
（3）配置 Shiro 过滤器工厂。
（4）开启对 Shiro 注解的支持。

```
Shiro 配置类需要的依赖
-----------------------------------------------------------
    <!-- 登录验证框架的依赖 -->
    <dependency>
        <groupId>org.apache.shiro</groupId>
        <artifactId>shiro-spring</artifactId>
        <version>1.3.2</version>
    </dependency>
    <!-- shiro-redis -->
    <dependency>
        <groupId>org.crazycake</groupId>
        <artifactId>shiro-redis</artifactId>
        <version>2.4.2.1-RELEASE</version>
    </dependency>
    <dependency>
```

```xml
    <groupId>org.apache.shiro</groupId>
    <artifactId>shiro-core</artifactId>
    <version>1.3.2</version>
</dependency>
<dependency>
    <groupId>org.apache.shiro</groupId>
    <artifactId>shiro-web</artifactId>
    <version>1.3.2</version>
</dependency>
```

1. 创建数据源 CustomRealm

CustomRealm 是 Shiro 进行登录或权限校验的逻辑核心，它继承了抽象类 AuthorizingRealm，需要重写 3 个方法，分别是 supports、doGetAuthenticationInfo 和 doGetAuthorizationInfo。supports 的作用是为了让 CustomRealm 支持 JWT 的凭证校验，doGetAuthenticationInfo 的作用是登录认证校验，doGetAuthorizationInfo 的作用是进行权限校验。CustomRealm 类的具体实现见 7.6.4 节和 7.6.5 节。

2. 创建安全管理器

安全管理器的主要功能和配置流程如下：

（1）SecurityManager 管理所有的 Realm，通过对 CustomRealm 的注入获取 CustomRealm 认证授权信息。

（2）注入 SessionManager，提供一个应用程序所需的所有企业级会话管理。

（3）注入 CacheManager，在 CacheManager 中注入 RedisManager，将 Redis 进行整合，用来处理和配置资源缓存，以提高系统性能。

（4）在 DefaultWebSecurityManager 中关闭 Shiro 自带的 Session，这样用户就不能通过 Session 方式登录 Shiro，而是采用 JWT 凭证方式登录。

（5）在 com.imooc.demo.shiro 包中创建 SessionConfig 类，定义获取 getSessionId 的方法，再通过 SessionManager 方法将 RedisSessionDAO 注入 SessionConfig 对象中，进而完成 Redis 对用户授权信息的管理。需要注意的是，在注入 Redis 后，用户只要身份验证和授权成功，再访问资源时就只需进行身份验证，从而大大减轻了系统请求压力，提高了系统整体性能。

ShiroConfig 类的 SecurityManager 方法文件代码

```java
import com.imooc.demo.dao.UserMapper;
import com.imooc.demo.shiro.SessionConfig;
import org.apache.shiro.mgt.DefaultSessionStorageEvaluator;
import org.apache.shiro.mgt.DefaultSubjectDAO;
import org.apache.shiro.mgt.SecurityManager;
import org.apache.shiro.session.mgt.SessionManager;
import org.apache.shiro.spring.LifecycleBeanPostProcessor;
import org.apache.shiro.spring.security.interceptor.AuthorizationAttributeSourceAdvisor;
import org.apache.shiro.spring.web.ShiroFilterFactoryBean;
import org.apache.shiro.web.mgt.DefaultWebSecurityManager;
import org.crazycake.shiro.RedisCacheManager;
import org.crazycake.shiro.RedisManager;
import org.crazycake.shiro.RedisSessionDAO;
import org.springframework.aop.framework.autoproxy.DefaultAdvisorAutoProxyCreator;
import org.springframework.context.annotation.Bean;
```

```java
import org.springframework.context.annotation.Configuration;
import com.imooc.demo.filter.JWTFilter;
import com.imooc.demo.shiro.CustomRealm;
import javax.servlet.Filter;
import java.util.HashMap;
import java.util.Map;
@Configuration
public class ShiroConfig {
    /**
     * 注入 SecurityManager
     */
    @Bean
    public SecurityManager securityManager(CustomRealm customRealm) {
        DefaultWebSecurityManager securityManager = new DefaultWebSecurityManager();
        //设置自定义 Realm
        securityManager.setRealm(customRealm);
        securityManager.setSessionManager(sessionManager());
        securityManager.setCacheManager(cacheManager());
        /*
         * 关闭 Shiro 自带的 Session
         */
        DefaultSubjectDAO subjectDAO = new DefaultSubjectDAO();
        DefaultSessionStorageEvaluator defaultSessionStorageEvaluator =
                        new DefaultSessionStorageEvaluator();
        defaultSessionStorageEvaluator.setSessionStorageEnabled(false);
        subjectDAO.setSessionStorageEvaluator(defaultSessionStorageEvaluator);
        securityManager.setSubjectDAO(subjectDAO);
        return securityManager;
    }
```

RedisManager 方法文件代码

```java
    /**
     * 配置 Shiro RedisManager
     * 使用的是 Shiro-Redis 开源插件
     * RedisManager 类默认参数：
     *     private String host = "127.0.0.1";
     *     private int port = 6379;
     *     private int expire = 0;
     *     private int timeout = 0;
     *     private String password = "";
     *     private static JedisPool jedisPool = null;
     * 可在方法中重置参数
     * @return
     */
    public RedisManager redisManager() {
        RedisManager redisManager = new RedisManager();
        redisManager.setHost("127.0.0.1");
        redisManager.setPort(6379);
        redisManager.setExpire(14400);
        redisManager.setTimeout(5000);
        redisManager.setPassword("");
        return redisManager;
    }
```

CacheManager 方法文件代码

```java
/**
 * CacheManager 缓存 Redis 实现
 * 使用的是 Shiro-Redis 开源插件 *
 * @return
 */
public RedisCacheManager cacheManager() {
    RedisCacheManager redisCacheManager = new RedisCacheManager();
    redisCacheManager.setRedisManager(redisManager());
    return redisCacheManager;
}
```

RedisSessionDAO 方法文件代码

```java
/**
 * 通过 Redis 实现 RedisSessionDAO Shiro SessionDao 层
 * 使用的是 Shiro-Redis 开源插件
 */
@Bean
public RedisSessionDAO redisSessionDAO() {
    RedisSessionDAO redisSessionDAO = new RedisSessionDAO();
    redisSessionDAO.setRedisManager(redisManager());
    return redisSessionDAO;
}
```

SessionManager 方法文件代码

```java
//自定义 SessionManager
@Bean
public SessionManager sessionManager() {
    SessionConfig mySessionManager = new SessionConfig();
    mySessionManager.setSessionDAO(redisSessionDAO());
    return mySessionManager;
}
```

SessionConfig.java 类文件代码

```java
package com.imooc.demo.shiro;
import org.apache.shiro.web.session.mgt.DefaultWebSessionManager;
import org.apache.shiro.web.servlet.ShiroHttpServletRequest;
import org.apache.shiro.web.util.WebUtils;
import org.springframework.util.StringUtils;
import javax.servlet.ServletRequest;
import javax.servlet.ServletResponse;
import java.io.Serializable;
public class SessionConfig extends DefaultWebSessionManager {
    public SessionConfig() {
        super();
    }
    @Override
    protected Serializable getSessionId(ServletRequest request, ServletResponse response) {
        String id = WebUtils.toHttp(request).getHeader("token");
        if (!StringUtils.isEmpty(id) && !"null".equals(id)) {
            request.setAttribute(ShiroHttpServletRequest.REFERENCED_SESSION_ID_SOURCE, "Stateless request");
            request.setAttribute(ShiroHttpServletRequest.REFERENCED_SESSION_ID, id);
```

```
                request.setAttribute(ShiroHttpServletRequest.REFERENCED_SESSION_ID_IS_VALID,
Boolean.TRUE);
                return id;
            } else {
                //否则按默认规则从 Cookie 取 SessionId
                return super.getSessionId(request, response);
            }
        }
    }
```

3. 配置 Shiro 过滤器工厂

在 Web 程序中,Shiro 进行权限控制时通过一组过滤器集合进行操作。过滤器配置需要有以下 4 个步骤:

(1) 创建过滤器工厂;
(2) 设置安全管理器;
(3) 通用配置(跳转登录页面,为授权跳转的页面);
(4) 设置过滤器集合。

需要强调的是,除访问 /unauthorized/** 和登录请求/login 不通过 JWTFilter 外,其他所有资源访问都需要自定义过滤器 JWTFilter 进行拦截。前端在设置 Axios 拦截器时,对 response 的 error 处理设计了对后端/unauthorized 的请求,以此获取请求资源中验证授权异常的捕获及显示。

ShiroConfig 类的 factory 方法文件代码

```
/**
 * 首先通过 filter 进行拦截,如果 filter 检测到请求头存在 Token,则用 Token 进行 login,
然后通过 Realm 进行验证
 */
@Bean
public ShiroFilterFactoryBean factory(SecurityManager securityManager, UserMapper userMapper) {
    ShiroFilterFactoryBean factoryBean = new ShiroFilterFactoryBean();
    //添加自己的过滤器并且取名为 jwt
    Map<String, Filter> filterMap = new HashMap<>();
    //设置自定义的 JWT 过滤器
    filterMap.put("jwt", new JWTFilter(userMapper));
    factoryBean.setFilters(filterMap);
    factoryBean.setSecurityManager(securityManager);
    //设置无权限时跳转的 URL
    factoryBean.setUnauthorizedUrl("/unauthorized/无权限");
    Map<String, String> filterRuleMap = new HashMap<>();
    //所有请求通过自己的 JWTFilter
    filterRuleMap.put("/**", "jwt");
    //访问 /unauthorized/** 不通过 JWTFilter
    filterRuleMap.put("/unauthorized/**", "anon");
    filterRuleMap.put("/login", "anon");
    factoryBean.setFilterChainDefinitionMap(filterRuleMap);
    return factoryBean;
}
```

4. 开启对 Shiro 注解的支持

DefaultAdvisorAutoProxyCreator 继承了 AbstractAdvisorAutoProxyCreator,是一种

基于 Advisor 的自动代理创建者的实现。切面 Advisor 是切点和增强的复合体。DefaultAdvisorAutoProxyCreator 能够扫描 Advisor，并将 Advisor 自动织入匹配的目标 Bean 中，即为匹配的目标 Bean 自动创建代理，进而实现 Spring AOP 切面编程。

ShiroConfig 类的 factory 方法文件代码

```
@Bean
    public DefaultAdvisorAutoProxyCreator defaultAdvisorAutoProxyCreator() {
        DefaultAdvisorAutoProxyCreator defaultAdvisorAutoProxyCreator =
                            new DefaultAdvisorAutoProxyCreator();
        //强制使用cglib，防止重复代理和可能引起代理出错的问题
        defaultAdvisorAutoProxyCreator.setProxyTargetClass(true);
        return defaultAdvisorAutoProxyCreator;
    }
```

7.5 基于 Shiro 框架的用户登录设计

前后端分离项目的后端登录设计和前后端一体项目登录设计是不一样的。在进行前后端一体设计时，系统在用户登录成功后就会把用户信息放到 Session 中，供前端和后端使用。而在前后端分离项目中，因为不使用 Session，所以前后端要进行交互只能通过 JWT 格式的令牌，即通过 Token 值进行身份和授权认证，用户每次访问后端 Controller 资源都需要携带此令牌，以达到安全访问的目的。

后端的用户登录就是通过输入正确的用户名和密码后会得到一个 Token 值，以备用户访问其他 Controller 资源时使用。

7.5.1 用户登录的实体类设计

在登录验证过程中，需要用到几个实体类来完成用户登录。

1. 用户实体类 User

第 3 章已经介绍了表数据结构和创建数据库表的内容，下面将使用 MyBatis 的 ORM（Object Relational Mapping，对象关系映射）设计持久化类与表的映射关系。ORM 框架在运行时才能参照映射文件（XML）的信息，把对象持久化到数据库中，完成对数据库增、删、改、查的操作。

在项目 Demo 的 com.imooc.demo.entity 包中创建实体类 User。

User.java 文件代码

```
package com.imooc.demo.entity;
public class User {
    private String username;                    //用户名
    private String password;                    //密码
    private String permission;                  //权限
    private Integer ban;                        //是否封号
    private Role role;                          //角色
    private String note;                        //用户描述
    private String autograph;                   //电子签名
        //省略 getter 和 setter 方法
}
```

在资源文件夹 resources 目录下，创建 userMapper.xml 配置文件，设计 User 类的对象关系映射文件，其中 role 属性的 association 标签用于实现链表查询，使得 User 类的 role 属性和 Role 对象建立 1 对 1 的关联。映射文件标签类型为 resultMap，标签 id 为 UserResult。

User 类的 resultMap 标签文件代码

```xml
<resultMap type="com.imooc.demo.entity.User" id="UserResult">
    <id property="username" column="username"/>
    <result property="password" column="password"/>
    <result property="ban" column="ban"/>
    <result property="permission" column="permission"/>
    <result property="note" column="note"/>
    <result property="autograph" column="autograph"/>
    <association column="role" property="role" javaType="com.imooc.demo.entity.Role" resultMap="RoleResult"></association>
</resultMap>
```

2. 角色实体类 Role

数据库表 user 和 role 表存在主外键关联关系，对应到实体类 User 的属性 role 是角色 Role 类型。

在项目 Demo 的 com.imooc.demo.entity 包中创建实体类 Role。

Role.java 文件代码

```java
package com.imooc.demo.entity;
public class Role {
    private String role;            //角色名
    private String permissions;     //角色权限
    private String descr;           //角色描述
    //省略 getter 和 setter 方法
}
```

在资源文件夹 resources 目录下的 userMapper.xml 配置文件中，创建 Role 类的对象关系映射文件，标签类型为 resultMap，标签 id 为 RoleResult。

Role 类的 resultMap 标签文件代码

```xml
<resultMap type="com.imooc.demo.entity.Role" id="RoleResult">
    <id property="role" column="role"/>
    <result property="permissions" column="permissions"/>
    <result property="descr" column="descr"/>
</resultMap>
```

3. 接口访问返回对象封装类 ResultMap

为规范接口访问返回的数据格式，一般情况下，前后端分离项目需要设计一个数据返回的封装类，该类继承 HashMap<String, Object>类，返回的数据类型为键值对格式。ResultMap 中包含 success()方法、fail()方法、code()方法和 message()方法，分别代表访问成功、访问失败、响应代码和返回数据。

在项目 Demo 的 com.imooc.demo.shiro 包中创建实体类 ResultMap。

ResultMap.java 文件代码

```java
package com.imooc.demo.shiro;
```

```java
import org.springframework.stereotype.Component;
import java.util.HashMap;
@Component
public class ResultMap extends HashMap<String, Object> {
    public ResultMap() {
    }
    public ResultMap success() {
        this.put("result", "success");
        return this;
    }
    public ResultMap fail() {
        this.put("result", "fail");
        return this;
    }
    public ResultMap code(int code) {
        this.put("code", code);
        return this;
    }
    public ResultMap message(Object message) {
        this.put("message", message);
        return this;
    }
}
```

7.5.2 用户登录设计

从后端系统架构思想分析得知,除用户登录请求/login 等个别资源外,其他 Controller 资源都是需要授权验证的。本节通过用户登录的请求和 Shiro 身份验证的流程,设计展示 Shiro 框架强大且灵活的安全策略。用户登录设计思路为通过需求分析如何得到实现结果,进而结合 Shiro 框架进行实现。

1. 实现 Controller 控制层处理类

在 Demo 项目中创建 com.imooc.demo.web 包,用于客户端登录统一管理。在 Web 包下创建登录页处理类 LoginController,并编写登录逻辑代码。

LoginController.java 文件代码

```java
package com.imooc.demo.web;
import org.springframework.beans.factory.annotation.Autowired;
import org.springframework.web.bind.annotation.*;
import com.imooc.demo.dao.UserMapper;
import com.imooc.demo.entity.User;
import com.imooc.demo.service.UserService;
import com.imooc.demo.shiro.ResultMap;
import com.imooc.demo.util.JWTUtil;
import com.imooc.demo.util.MD5Util;
import java.io.UnsupportedEncodingException;
/**
 * 用户登录
 */
@RestController
public class LoginController {
    private final UserMapper userMapper;
```

```java
        private final ResultMap resultMap;

        @Autowired
        private UserService userService;
        @Autowired
        private RedisTemplate redisTemplate;
        @Autowired
        public LoginController(UserMapper userMapper, ResultMap resultMap) {
            this.userMapper = userMapper;
            this.resultMap = resultMap;
        }
        @PostMapping("/login")
        public ResultMap login(@RequestParam("username") String username,
                               @RequestParam("password") String password) {
            User user = userService.findByUsername(username);
            String pwd = MD5Util.toMD5(password);
            if (user == null) {
                return resultMap.fail().code(400).message("用户名错误");
            } else {
                String realPassword = user.getPassword();
                int ban = user.getBan();
                if (!realPassword.equals(pwd)) {
                    return resultMap.fail().code(401).message("密码错误");
                }
                if (ban == 1) {
                    return resultMap.fail().code(402).message("账户被封号");
                }
            redisTemplate.opsForValue().set(username, JWTUtil.createToken(username));
            return resultMap.success().code(200).message(JWTUtil.createToken(username));
            }
        }
```

在上述代码文件中,/login 是用户登录的请求地址,请求方法为 POST。从理论上说,POST 请求比 GET 请求要更安全。因为一般的网页请求基本都是 GET 方法,所以使用请求可以避免很多不必要的访问,安全性也会大大提高。POST 方法的请求逻辑如下。

(1) 在该类中注入 UserService 接口,调用其对象的实现类 UserServiceImpl 的 findByUsername 方法,通过登录用户传入的用户名从数据库查找是否有该用户,如果存在该用户则生成一个 User 对象。

(2) 将用户传入的密码 password 用 MD5 工具类加密,与用户对象中的密码,即数据库中的 password 进行比对。

(3) 判断结果分为以下 4 种情况。

第 1 种情况:如果用户对象为空,即通过用户名没有查到用户对象,则说明用户名错误。通过封装的结果类 resultMap 返回给前端,消息为"用户名错误",代码为 400。

第 2 种情况:如果用户对象不为空,但用户传入的密码和数据库的密码不一致,则说明密码错误。通过封装的结果类 resultMap 返回给前端,消息为"密码错误",代码为 401。

第 3 种情况:如果用户对象不为空,并且用户传入的密码和数据库的密码一致,但查询数据库 user 表得到的 ban 字段为 1,则说明账户被封号。通过封装的结果类 resultMap 返回给前端,消息为"账户被封号",代码为 402。

第 4 种情况：如果用户对象不为空，并且用户传入的密码和数据库的密码一致，查询数据库 user 表得到的 ban 字段为 0，则说明是合法用户。通过封装的结果类 resultMap 返回给前端一个 message，该 message 是由工具类的 JWTUtil.createToken(username) 方法生成的 JWT 格式的 Token 值，即身份认证令牌，代码为 200。

(4) 在该类中注入 RedisTemplate 接口，用户访问成功后将生成的 Token 存入 Redis，key 为 username，value 为 Token 值，用于 Shiro 的身份认证和用户退出登录等操作。

2. 生成 JWT 格式的 token 的工具类 JWTUtil

在 Demo 项目中创建 com.imooc.demo.util 包，在包中创建工具类 JWTUtil，用于生成 JWT 格式的 Token 值。

JWTUtil.java 文件代码

```java
package com.imooc.demo.util;
import com.auth0.jwt.JWT;
import com.auth0.jwt.JWTVerifier;
import com.auth0.jwt.algorithms.Algorithm;
import com.auth0.jwt.exceptions.JWTDecodeException;
import com.auth0.jwt.interfaces.DecodedJWT;
import java.io.UnsupportedEncodingException;
import java.util.Date;
public class JWTUtil {
    //过期时间 24 小时
    private static final long EXPIRE_TIME = 60 * 24 * 60 * 1000;
    //密钥
    private static final String SECRET = "SHIRO";
    /**
     * @param username 用户名
     * @return 加密的 Token
     */
    public static String createToken(String username) {
        try {
            Date date = new Date(System.currentTimeMillis() + EXPIRE_TIME);
            Algorithm algorithm = Algorithm.HMAC256(SECRET);
            //附带 username 信息
            return JWT.create()
                    .withClaim("username", username)
                    //到期时间
                    .withExpiresAt(date)
                    //创建一个新的 JWT，并使用给定的算法进行标记
                    .sign(algorithm);
        } catch (UnsupportedEncodingException e) {
            return null;
        }
    }
}
```

生成的 token 值由 3 个部分组成，里面包含了用户名、密钥和过期时间。密钥（SECRET）由 HMAC256 方法进行加密处理，最后会经过 Base64 加密方式进行编码。SECRET 就是在最后第二次加密时加的盐，算是一个密钥（只保留在服务器），不透露给前端。前端用户通过在请求头 Header 中携带该 Token 值，访问其他 Controller 资源，以达到认证

授权的目的。

7.5.3 项目的启动类 DemoApplication

Spring Boot 项目由启动类启动，Demo 项目的启动类 DemoApplication 是在创建项目成功后自动生成的，类中需要有程序入口 main() 方法。启动类上必须注入 @SpringBootApplication 注解才可启动项目。

DemoApplication.java 文件代码

```
package com.imooc.demo;
import org.mybatis.spring.annotation.MapperScan;
import org.springframework.boot.SpringApplication;
import org.springframework.boot.autoconfigure.SpringBootApplication;
import org.springframework.context.annotation.Bean;
import org.springframework.context.support.PropertySourcesPlaceholderConfigurer;
@SpringBootApplication
@MapperScan(value = "com.imooc.demo.dao")
public class DemoApplication {
    public static void main(String[] args) {
        SpringApplication.run(DemoApplication.class, args);
    }
}
```

7.5.4 项目的启动测试

在 IDEA 开发界面选中 DemoApplication，单击右侧的启动按钮即可启动，启动界面如图 7-12 所示，启动成功效果如图 7-13 所示。

由于 Spring Boot 项目封装了 Tomcat，所以项目启动后直接访问地址即可，不用再另外配置 Tomcat。

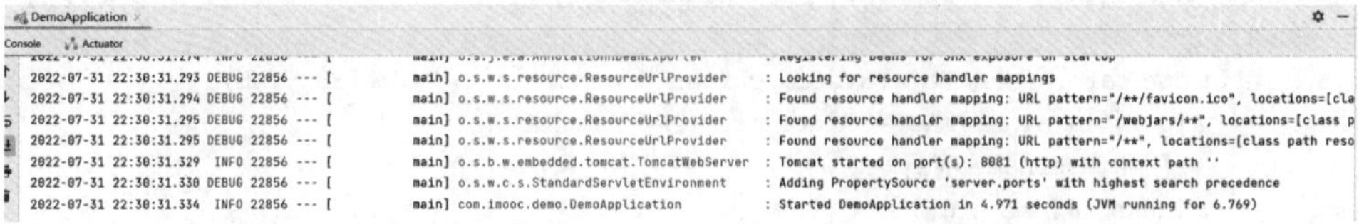

图 7-12　IDEA 项目启动界面

图 7-13　IDEA 项目启动成功界面

本项目使用接口测试工具 Postman 进行接口资源访问测试。Postman 相当于一个客户端，它可以模拟用户发起的各类 HTTP 请求，将请求数据发送至服务端，获取对应的响应结果，从而验证响应中的结果数据是否和预期值相匹配，并确保开发人员能够及时处理接口中的 bug，进而保证产品上线之后的稳定性和安全性。Postman 与浏览器的区别在于有的浏览器不能输出 JSON 格式，而 Postman 能更直观地体现接口返回的结果。

打开 Postman 测试工具，在地址栏中输入 127.0.0.1:8081/login，请求方法选择 POST。设置 param 参数 1 的 key 为 username，value 为 liuzh；设置 param 参数 2 的 key 为 password，value 为 123456。单击 send 按钮，访问成功后的结果如图 7-14 所示。

输出格式即为结果封装类 ResultMap 格式，其中输出的 message 长字符串为 JWT 格式的 Token 值。

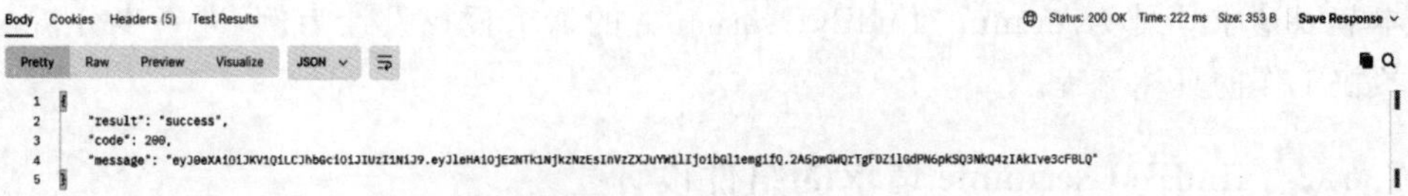

图 7-14 请求/login 接口访问成功结果图

7.6 Apache Shiro 认证授权安全框架设计

从 7.4 节 Shiro 工作机制的授权流程分析可知,用户登录成功后需要根据角色权限访问相应的 Controller 资源,以达到用户角色权限分配的目的。基于此,下面对 Shiro 认证授权安全框架进行设计。

7.6.1 Shiro 的认证授权工作流程

根据 Shiro 框架的运行机制,设计出 Shiro 框架认证授权过程中各组件的执行顺序和执行任务,进而得出具体认证授权工作执行流程图。Shiro 认证授权工作执行流程图如图 7-15 所示。

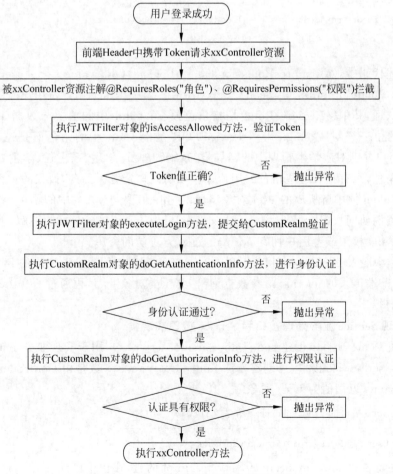

图 7-15 Shiro 认证授权工作执行流程图

本节以请求地址为 /admin /findByUsername 的请求授权设计为例，进一步分析 Shiro 框架授权工作的具体流程。

7.6.2 findByUsername 请求的组件设计

1. 实现 Controller 控制层处理类

在 Demo 项目的 com.imooc.demo.web 包中创建用户管理类 AdminController，新建 findByUsername 方法。

findByUsername 方法文件代码

```
@RestController
@RequestMapping("/admin")                                    //窄化请求路径
public class AdminController {
@Autowired
private UserService userService;
@GetMapping("/findByUsername")                               //对前端发起的 GET 请求进行拦截
@RequiresPermissions("normal")
@RequiresRoles(logical = Logical.OR, value = { "user", "admin", "student", "depp", "stum",
"dean" })
    public ResultMap findByUsername(String username) {
        User user = userService.findByUsername(username);
        return resultMap.success().code(200).message(user);
    }
}
```

在上述代码中，@RequiresRoles(logical = Logical.OR，value = { "user"，"admin"，"student"，"depp"，"stum"，"dean" })是对 token 值中携带的用户及用户所属角色权限进行拦截和认证，其中@RequiresRoles 是指角色授权；value 中的值要与数据库 role 表中的 role 字段的值一一对应；logical = Logical.OR 是指 value 中的角色值是逻辑或的关系；6 类角色用户分别对应学生管理中的辅导员、系统管理员、学生、系主任、学生处处长和教务处处长等角色，只有拥有这些角色的用户才可以访问该资源。因此，授权的过程就是验证用户请求的 token 值中解析出的 user 对象属性的 role 值要与 value 中的哪个值匹配，方可授予对应的资源访问权限。另外，添加@RequiresPermissions("normal")注解，就需要用户角色拥有 normal 权限或者用户拥有 normal 权限，方可访问资源。

在该类中注入 UserService 接口对象 userService，通过调用 userService 的 findByUsername 方法，根据前端传来的 username 参数查询出 User 类对象 user，再通过 resultMap 结果类封装后传给前端。

2. 实现 Service 业务处理接口和 Service 接口实现类

（1）在 Demo 项目中创建 com.imooc.demo.service 包，用于实现业务逻辑处理。在该包下创建 UserService 接口，再创建方法 findByUsername，参数是由 Controller 层传来的用户名 username，返回值为 User 对象。

UserService 接口的 findByUsername 方法文件代码

```
public User findByUsername(String username);
```

（2）创建名为 com.imooc.demo.service.impl 的包，在包中创建 UserServiceImpl 实现

类,实现 UserService 接口。该类需添加@Service 和@Transactional 注解,添加@Service 注解使 UserServiceImpl 实现类纳入 Spring 容器,添加@Transactional 注解使事务管理生效,尤其是对数据库增、删、改的操作,否则将不能对数据库进行有效处理。该实现类注入了 Mapper 接口的 UserMapper 对象,调用 findByUsername 方法。

UserServiceImpl 类的 findByUsername 方法文件代码

```
package com.imooc.demo.service.Impl;
import org.springframework.beans.factory.annotation.Autowired;
import org.springframework.stereotype.Service;
import org.springframework.transaction.annotation.Transactional;
import com.imooc.demo.dao.UserMapper;
import com.imooc.demo.entity.User;
import com.imooc.demo.service.UserService;
@Service
@Transactional
public class UserServiceImpl implements UserService{
//注入 Mapper 接口
@Autowired
    private UserMapper usermapper;
@Override
    public User findByUsername(String username) {
        return usermapper.findUserByName(username);
    }
}
```

3. 设计 DAO 接口和 userMapper.xml 文件,实现对数据库的操作

(1) 在 Demo 项目中创建 com.imooc.demo.dao 包,创建 UserMapper 接口,再创建方法 findByUsername,参数是由 Service 层传来的用户名 username,返回值为 User 对象。

UserMapper.java 文件代码

```
User findUserByName(String username);
```

(2) 在资源文件夹 resources 目录下的 userMapper.xml 配置文件中创建 id 为 findByUsername 的 SQL 语句,对数据库查询的结果标签类型为 resultMap,id 为 UserResult 的嵌套对象数据,实现 User 和 Role 对象的联合查询。

findByUsername 的 SQL 语句文件代码

```
<select id="findByUsername" parameterType="String"
    resultMap="UserResult">
    SELECT u.username, u.password, u.ban,
    u.permission,u.autograph,u.note,r.role,r.descr FROM
    user u left join role r on
    u.role = r.role where u.username = #{username}
</select>
```

7.6.3　JWTFilter 类对 token 值的过滤设计

在项目 Demo 中创建 com.imooc.demo.filter 包,在包中创建 JWTFilter 类,该类继承了 BasicHttpAuthenticationFilter 类,并重写父类的 isAccessAllowed 方法、isLoginAttempt

方法和 executeLogin 方法。在重写方法中，isAccessAllowed 是主方法，该方法通过调用 isLoginAttempt 方法判断 Header 中是否携带 Token，如果携带 Token 就执行 executeLogin 方法，最后将 Token 交给 CustomRealm 进行身份验证和授权验证，具体验证流程如下。

（1）执行 executeLogin 方法，如果身份验证和授权验证成功，则返回 true，访问 Controller 接口。

（2）执行 executeLogin 方法，如果身份验证和授权验证失败，则通过执行 executeLogin 方法获取 CustomRealm 中认证和授权的异常错误信息 e.getMessage()，再通过 responseError 方法将异常错误信息写入数据库，最后提供给 /unauthorized 使用。

isAccessAllowed 方法文件代码

```java
package com.imooc.demo.filter;
import org.apache.shiro.authz.UnauthorizedException;
import org.apache.shiro.web.filter.authc.BasicHttpAuthenticationFilter;
import org.slf4j.Logger;
import org.slf4j.LoggerFactory;
import org.springframework.http.HttpStatus;
import org.springframework.web.bind.annotation.RequestMethod;
import com.imooc.demo.shiro.JWTToken;
import javax.servlet.ServletRequest;
import javax.servlet.ServletResponse;
import javax.servlet.http.HttpServletRequest;
import javax.servlet.http.HttpServletResponse;
import java.io.IOException;
import java.net.URLEncoder;
public class JWTFilter extends BasicHttpAuthenticationFilter {
    private Logger logger = LoggerFactory.getLogger(this.getClass());
    @Override
    protected boolean isAccessAllowed(ServletRequest request, ServletResponse response, Object mappedValue) throws UnauthorizedException {
        //判断请求的请求头是否带有 Token
        if (isLoginAttempt(request, response)) {
        //如果存在，则进入 executeLogin 方法执行登录，检查 Token 是否正确
            try {
                executeLogin(request, response);
                return true;
            } catch (Exception e) {
                //通过 e.getMessage() 获取 Token 错误信息
                responseError(response, e.getMessage());
            }
        }
        //如果请求头不存在 Token,则可能是执行登录操作或游客状态访问，无须检查 token,直接返回 true
        return true;
    }
}
```

isLoginAttempt 方法文件代码

```java
@Override
    protected boolean isLoginAttempt(ServletRequest request, ServletResponse response) {
        HttpServletRequest req = (HttpServletRequest) request;
        String token = req.getHeader("Token");
```

```
        return token != null;
    }
```
executeLogin 方法文件代码

```
@Override
    protected boolean executeLogin(ServletRequest request, ServletResponse response) throws
Exception {
        HttpServletRequest httpServletRequest = (HttpServletRequest) request;
        String token = httpServletRequest.getHeader("Token");
        JWTToken jwtToken = new JWTToken(token);
        //提交给 Realm 进行登录,如果错误则会抛出异常并进行捕获
        getSubject(request, response).login(jwtToken);
        //如果没有抛出异常则代表登录成功,返回 true
        return true;
    }
```
responseError 方法文件代码

```
    /**
     * 将非法请求跳转到 /unauthorized/
     */
    private void responseError(ServletResponse response, String message) {
        try {
    HttpServletResponse httpServletResponse = (HttpServletResponse) response;
            //将异常信息写入数据库,提供给/unauthorized 使用
            int row = userMapper.updateEMessage(message);
            httpServletResponse.sendRedirect("/unauthorized");
        } catch (IOException e) {
            logger.error(e.getMessage());
        }
    }
```

7.6.4 CustomRealm 类对当前登录用户身份验证设计

在 Demo 项目的 com.imooc.demo.shiro 包中创建 CustomRealm 类,该类继承了 AuthorizingRealm 抽象类,并重写父类的 supports 方法、doGetAuthenticationInfo 方法和 doGetAuthorizationInfo 方法。doGetAuthenticationInfo 方法用来对当前登录用户进行身份验证,具体验证流程如下。

(1) 重写父类的 supports 方法,instanceof 是 Java 的保留关键字,它的作用是测试它左边的对象是否为它右边的类的实例,并返回 boolean 的数据类型。其目的是检验 token 值是否为 JWTToken 类的实例,结果为真时程序才可以继续进行,否则抛出异常。

(2) 通过 authenticationToken.getCredentials 方法获取 token。

(3) 通过工具类 JWTUtil.getUsername 方法解析 token 值获得 username 信息。

(4) 根据 username 从 Redis 中获取 key 为 username 的 token 值_token。

(5) 如果_token 为空或通过 JWTUtil.verify(token,username)方法检验失败时,抛出 "token 认证失败"异常。

(6) 如果通过 CustomRealm 类中注入的 userMapper 接口对象方法 userMapper. getPassword(username)从数据库中没有查出 password,抛出"用户不存在"异常。

(7) 如果通过 CustomRealm 类中注入的 userMapper 接口对象方法 userMapper.

checkUserBanStatus(username) 从数据库中查出 ban 的值为 1 时,抛出"该用户已被封号"异常。

Supports 方法文件代码

```java
@Override
    public boolean supports(AuthenticationToken token) {
        return token instanceof JWTToken;
    }
```

doGetAuthenticationInfo 方法文件代码

```java
package com.imooc.demo.shiro;
import com.imooc.demo.dao.UserMapper;
import com.imooc.demo.util.JWTUtil;
import org.apache.shiro.authc.AuthenticationException;
import org.apache.shiro.authc.AuthenticationInfo;
import org.apache.shiro.authc.AuthenticationToken;
import org.apache.shiro.authc.SimpleAuthenticationInfo;
import org.apache.shiro.authz.AuthorizationInfo;
import org.apache.shiro.authz.SimpleAuthorizationInfo;
import org.apache.shiro.realm.AuthorizingRealm;
import org.apache.shiro.subject.PrincipalCollection;
import org.springframework.beans.factory.annotation.Autowired;
import org.springframework.stereotype.Component;
import java.util.HashSet;
import java.util.Set;
@Component
public class CustomRealm extends AuthorizingRealm {
    private final UserMapper userMapper;
    @Autowired
    public CustomRealm(UserMapper userMapper) {
        this.userMapper = userMapper;
    }
    /**
     * 必须重写此方法,不然会报错
     */
    @Override
    public boolean supports(AuthenticationToken token) {
        return token instanceof JWTToken;
    }
    /**
     * 默认使用此方法进行用户名正确与否验证,错误时抛出异常
     */
    @Override
    protected AuthenticationInfo doGetAuthenticationInfo(AuthenticationToken authenticationToken)
throws AuthenticationException {
String token = (String) authenticationToken.getCredentials();
//解密获得 username
String username = JWTUtil.getUsername(token);
//根据 username 从 Redis 中获取 Token
String _token = (String) redisTemplate.opsForValue().get(username);
//用 username 校验 Token 是否有效
if (_token == null || !JWTUtil.verify(token, username)) {
    throw new AuthenticationException("Token 认证失败!");
```

```
}
//用 username 和数据库进行对比
String password = userMapper.getPassword(username);
if (password == null) {
    throw new AuthenticationException("该用户不存在!");
}
int ban = userMapper.checkUserBanStatus(username);
if (ban == 1) {
    throw new AuthenticationException("该用户已被封号!");
}
return new SimpleAuthenticationInfo(token, token, "MyRealm");
}
```
JWTUtil.verify 方法文件代码

```
public static boolean verify(String token, String username) {
    try {
        Algorithm algorithm = Algorithm.HMAC256(SECRET);
        //在 Token 中附带了 username 信息
        JWTVerifier verifier = JWT.require(algorithm)
                .withClaim("username", username)
                .build();
        //验证 token
        verifier.verify(token);
        return true;
    } catch (Exception exception) {
        return false;
    }
}
```

JWTUtil.verify 方法通过 com.auth0.jwt.JWT.require 方法整合密钥 SECRET 信息，通过 withClaim 方法整合当前用户名信息，通过 build 方法生成一个 JWT 验证类 JWTVerifier 对象 verifier(系统当前时间信息默认加入)，最后通过 verifier.verify 方法验证 Token 值是否有效。Token 认证失败有以下几种情况：

(1) 用户名 username 不正确；

(2) 密钥 SECRET 不正确；

(3) Token 值超过有效期(本系统设定过期时间为 24h)；

(4) 用户已被注销。

7.6.5 CustomRealm 类对当前登录用户授权设计

CustomRealm 类中重写的 doGetAuthorizationInfo 方法对当前登录用户授权进行设计。doGetAuthorizationInfo 方法的授权认证流程如下：

(1) 通过 JWTUtil.getUsername(principals.toString()) 方法获取用户名信息。

(2) 生成一个授权信息 SimpleAuthorizationInfo 类对象 info。

(3) 通过用户名获取用户角色 role。

(4) 通过用户名获取用户角色拥有的权限信息 rolePermission。

(5) 通过用户名获取用户拥有的权限信息 permission。

(6) 将 role、permission、rolePermission 封装到 Set 作为 info.setRoles() 和 info.

setStringPermissions()的参数,将该参数添加到 SimpleAuthorizationInfo 类对象 info 中。

(7) 方法返回授权信息 SimpleAuthorizationInfo 类对象 info。

(8) 权限认证及异常抛出:通过用户名获取用户角色 role 与 Controller 注解 @RequiresRoles 的 value 值匹配,匹配成功则程序继续向下进行,匹配不成功则抛出 Subject does not have role [user]异常;通过用户名获取用户角色拥有的权限信息 rolePermission 及通过用户名获取用户拥有的权限信息 permission,与 user 表的 permission 字段值、role 表中的 permissions 字段值匹配,只要有一项能匹配成功即授权成功,没有匹配成功则抛出 Subject does not have permission [normal]异常。异常捕获通过 com.imooc.demo.handler 中创建的 GlobalExceptionHandler 类获取。

doGetAuthorizationInfo 方法文件代码

```java
/**
 * 只有当需要检测用户权限的时候才会调用此方法,例如 checkRole,checkPermission 之类的
 */
@Override
protected AuthorizationInfo doGetAuthorizationInfo(PrincipalCollection principals) {
    System.out.println("————CustomRealm 权限认证————");
    String username = JWTUtil.getUsername(principals.toString());
    SimpleAuthorizationInfo info = new SimpleAuthorizationInfo();
    //获得该用户角色
    String role = userMapper.getRole(username);
    //每个角色拥有默认的权限
    String rolePermission = userMapper.getRolePermission(username);
    //每个用户可以设置新的权限
    String permission = userMapper.getPermission(username);
    Set<String> roleSet = new HashSet<>();
    Set<String> permissionSet = new HashSet<>();
    //需要将 role 和 permission 封装到 Set 作为 info.setRoles()和 info.setStringPermissions() 的
    //参数
    roleSet.add(role);
    permissionSet.add(rolePermission);
    permissionSet.add(permission);
    //设置该用户拥有的角色和权限
    info.setRoles(roleSet);
    info.setStringPermissions(permissionSet);
    return info;
}
```

7.6.6 GlobalExceptionHandler 全局异常捕获设计

在 com.imooc.demo.handler 包中创建 GlobalExceptionHandler 类,在该类上添加 @RestControllerAdvice 注解和@Slf4j 注解,方法 handler 添加@ExceptionHandler 注解。这 3 个注解的主要作用如下:

(1) @RestControllerAdvice 是一个组合注解,由@ControllerAdvice 和@ResponseBody 组成,而@ControllerAdvice 继承了@Component,因此@RestControllerAdvice 本质上是 @Component,用于定义@ExceptionHandler、@InitBinder 和@ModelAttribute 方法,适用于所有使用@RequestMapping 的方法。

（2）@ExceptionHandler 注解用于指定异常处理方法。当与@RestControllerAdvice 配合使用时，用于全局异常处理。

（3）@Slf4j 注解用于日志管理。

GlobalExceptionHandler 类文件代码

```java
package com.imooc.demo.handler;
import com.imooc.demo.service.UserService;
import com.imooc.demo.shiro.ResultMap;
import lombok.extern.slf4j.Slf4j;
import org.apache.shiro.ShiroException;
import org.springframework.beans.factory.annotation.Autowired;
import org.springframework.web.bind.annotation.*;
@Slf4j
@RestControllerAdvice
public class GlobalExceptionHandler {
    private final ResultMap resultMap;
    @Autowired
    private UserService userService;
    public GlobalExceptionHandler(ResultMap resultMap) {
        this.resultMap = resultMap;
    }
    @ExceptionHandler(value = ShiroException.class)
    public ResultMap handler(ShiroException e) {
        log.error("Shiro 异常 ShiroException...{}", e.getMessage());
        return resultMap.fail().code(408).message(e.getMessage());
    }
    @ExceptionHandler(value = RuntimeException.class)
    public ResultMap handler(RuntimeException e) {
        log.error("运行时异常...{}", e);
        //异常信息写入数据库
        userService.updateEMessage(e.getMessage());
        return resultMap.fail().code(400).message(e.getMessage());
    }
}
```

7.7　Apache Shiro 认证授权测试用例

7.7.1　findByUsername 请求成功用例

findByUsername 请求成功的条件如下：

（1）用户已经登录成功，并已获得 Token 值；

（2）用户角色需要匹配成功注解@RequiresRoles 的 value 值的一项；

（3）@RequiresPermissions("normal")注解的 normal 值匹配成功用户角色的 permissions 值或用户的 permission 值。

认证请求时，在 Headers 的 key 中填入 Token，在 value 中填入用户登录成功获取的 message(Token 值)。请求成功后，返回 resultMap 类对象给前端，result 为 success，message 值

为 user 对象，code 为 200。用 Postman 进行测试，测试结果如下所示。

```
{
    "result": "success",
    "code": 200,
    "message": {
        "username": "liuzh",
        "password": "49ba59abbe56e057",
        "permission": "normal",
        "ban": 0,
        "role": {
            "role": "admin",
            "permissions": "vip",
            "descr": "系统管理员"
        },
        "note": "刘仲会",
        "autograph": "data:image/jpeg;base64,/9j/4AAQSkZJRgABAQEAYABgAAD/
        ......
    }
}
```

7.7.2　findByUsername 请求身份认证失败用例

根据 7.6.2 节的内容可知，findByUsername 请求过程是根据 Token 值先进行身份认证，再进行授权认证。这里先进行身份认证失败的用例测试，验证时将用户登录成功的 Token 值放入 Headers 中，设置 key 为 Token、value 为 Token 值。

身份认证失败用例测试的实现步骤如下。

（1）在验证 token 方法 verify 中，重新设置参数 username。例如，设置 username＝username＋'l'，重新启动项目，用 Postman 发送 127.0.0.1:8081/admin/findByUsername?username＝liuzh 请求，请求方法为 GET，测试结果如下。

```
{
    "result": "fail",
    "code": 409,
    "message": "token 认证失败"
}
```

（2）将 Headers 中的 Token 值清空，在 postman 中发送同样的请求，测试结果如下。

```
{
    "result": "fail",
    "code": 409,
    "message": "token 认证失败"
}
```

（3）修改登录的用户名 liuzh，如修改为 liuzhaa，在 postman 中发送同样的请求，测试结果如下。

```
{
    "result": "fail",
    "code": 409,
    "message": "用户不存在!"
}
```

(4) 将数据库 user 表中的 liuzh 用户的 ban 值由 0 修改为 1，在 postman 中发送同样的请求，测试结果如下。

```
{
    "result": "fail",
    "code": 409,
    "message": "该用户被封号！"
}
```

(5) 将用户 liuzh 注销，在 postman 中访问 127.0.0.1:8081/admin/findRoutesByRole?role=admin&username=liuzh 接口，Headers 中放入已注销用户 liuzh 登录成功获得的 Token 值，发送请求，测试结果如下。

```
{
    "result": "fail",
    "code": 409,
    "message": "token 认证失败"
}
```

在上述测试过程中，通过 JWTFilter 类中的 responseError 方法跳转到 /unauthorized 资源控制器，并通过 String message = userService.getEMessage(1)语句从数据库获取异常信息，最后通过 return resultMap.fail().code(409).message(message)返回给前端。

```
@RequestMapping(path = "/unauthorized")
public ResultMap unauthorized() {
    String message = userService.getEMessage(1);
    return resultMap.success().code(409).message(message);
}
```

7.7.3 findByUsername 请求授权认证失败用例

授权认证失败用例测试的实现步骤如下。

(1) 在 findByUsername 方法的@RequiresRoles 注解中，将 value 值的 admin 角色删去。重新启动项目，发送同样的请求，用 Postman 进行测试，测试结果如下。

```
{
    "result": "fail",
    "code": 408,
    "message": "Subject does not have role [user]"
}
```

(2) 在 findByUsername 方法的@RequiresPermissions("normal")注解中，令 normal 值与数据库 user 表的 permission 值、role 表的 permissions 值不一致，如修改为@RequiresPermissions("normalaa")。重新启动项目，发送同样的请求，用 postman 进行测试，测试结果如下。

```
{
    "result": "fail",
    "code": 408,
    "message": "Subject does not have permission [normalaa]"
}
```

在上述测试过程中，通过 GlobalExceptionHandler 类中的 handler(ShiroException e)

方法获取到异常信息 e.getMessage()，最后通过 return resultMap.fail().code(408).message(e.getMessage())返回给前端。

handler 方法文件代码

```
@ExceptionHandler(value = ShiroException.class)
public ResultMap handler(ShiroException e) {
    log.error("Shiro 异常 ShiroException...{}", e.getMessage());
    return resultMap.fail().code(408).message(e.getMessage());
}
```

7.7.4 用户授权 Redis 缓存管理测试

该项目通过添加 Shiro-Redis 依赖，注入 Redis 实现了 redisManager 缓存管理。Redis 是一个高性能的 key-value 数据库，其性能极高。Redis 读的速度是 110000 次/s，写的速度是 81000 次/s。Redis 支持数据的持久化，可以将内存中的数据保存在磁盘中，重启时可以再次加载进行使用。本测试对 redisManager 如何实现缓存管理及提高系统抗压性能进行说明，并对如何实现 redisManager 缓存管理进行介绍。

Redis 缓存管理测试的实现步骤如下。

(1) 安装 Redis。

将安装包 Redis-x64-3.2.100 放到系统的某个路径，如 D:\IDEA\，双击安装包 Redis-x64-3.2.100 中的 Redis-server.exe 即可启动 Redis，如图 7-16 所示。

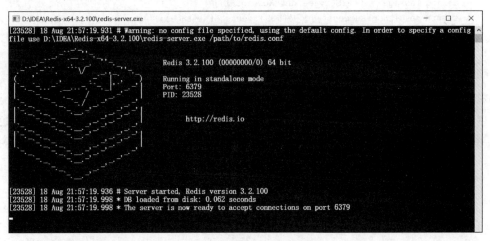

图 7-16 Redis 启动成功界面

双击 redis-cli.exe 打开客户端，Redis 安装默认地址为本机地址 127.0.0.1，默认端为 6379，默认密码为空。在客户端中不用输入密码，直接进行查看即可。在启用 Redis 前，输入命令 keys *，数据库显示没有数据，如图 7-17 所示。

图 7-17 Redis 数据库的初始数据

(2) 在 shiro 没有整合 redis 时，测试 shiro 的认证和授权情况。

在 ShiroConfig 配置类中的 securityManager 方法中，将 securityManager.setCacheManager(cacheManager()) 语句删除，即取消缓存管理的注入，用 postman 发送

127.0.0.1:8081/admin/findByUsername?username=liuzh 请求，查看 IDEA 的控制台输出，如图 7-18 所示。

```
——CustomRealm身份认证方法——
2022-08-19 19:59:54.160  INFO 24812 --- [nio-8081-exec-1] c.m.v.c.i.AbstractPoolBackedDataSource
2022-08-19 19:59:54.418 DEBUG 24812 --- [nio-8081-exec-1] o.s.web.servlet.DispatcherServlet
2022-08-19 19:59:54.421 DEBUG 24812 --- [nio-8081-exec-1] s.w.s.m.m.a.RequestMappingHandlerMapping
2022-08-19 19:59:54.424 DEBUG 24812 --- [nio-8081-exec-1] s.w.s.m.m.a.RequestMappingHandlerMapping
2022-08-19 19:59:54.424 DEBUG 24812 --- [nio-8081-exec-1] o.s.web.servlet.DispatcherServlet
——CustomRealm权限认证——
```

图 7-18　没有添加 redis 缓存时访问接口的 IDEA 控制台输出

由图 7-18 可以看出，本次请求既执行了 doGetAuthenticationInfo 方法进行身份认证，又执行了 doGetAuthorizationInfo 方法进行权限认证。

（3）在 shiro 整合 redis 后，测试 shiro 的认证和授权情况。

在 ShiroConfig 配置类中的 securityManager 方法中，添加 securityManager.setCacheManager(cacheManager()) 语句，即注入缓存管理，重新启动项目，用 postman 发送 127.0.0.1:8081/admin/findByUsername?username=liuzh 请求，查看 IDEA 的控制台输出，如图 7-19 所示。

```
——CustomRealm身份认证方法——
2022-08-19 20:20:41.569 DEBUG 12616 --- [nio-8081-exec-6] o.s.web.servlet.DispatcherServlet
2022-08-19 20:20:41.569 DEBUG 12616 --- [nio-8081-exec-6] s.w.s.m.m.a.RequestMappingHandlerMapping
2022-08-19 20:20:41.569 DEBUG 12616 --- [nio-8081-exec-6] s.w.s.m.m.a.RequestMappingHandlerMapping
2022-08-19 20:20:41.569 DEBUG 12616 --- [nio-8081-exec-6] o.s.web.servlet.DispatcherServlet
——CustomRealm权限认证——
```

图 7-19　添加 redis 缓存后首次访问接口的 IDEA 控制台输出

由图 7-19 可以看出，输出结果与没有注入缓存管理时是相同的。现在打开客户端 redis-cli.exe，输入 keys *，发现 Redis 数据库中已经写入了用户的授权信息，如图 7-20 所示。

```
D:\IDEA\Redis-x64-3.2.100\redis-cli.exe
127.0.0.1:6379> keys *
1) "\xac\xed\x00\x05sr\x002org.apache.shiro.subject.SimplePrincipalCollection\xa8\x7fX%\xc6\xa3\bJ\x03\x00\x01L\x00\x0f
realmPrincipalst\x00\x0fLjava/util/Map;xpsr\x00\x17java.util.LinkedHashMap4\xc0N\\\x101\xc0\xfb\x02\x00\x01Z\x00\x0bacce
ssOrderxr\x00\x11java.util.HashMap\x05\a\xda\xc1\xc3\x16\xd1\x03\x00\x02F\x00\nloadFactorI\x00\tthresholdxp?@\x00\x00\x00
\x00\x00\x0cw\b\x00\x00\x10\x00\x00\x00\x01t\x00\aMyRealmsr\x00\x17java.util.LinkedHashSet\xd8\xd7Z\x95\xdd*\x1e
\x02\x00\x00xr\x00\x11java.util.HashSet\xbaD\x85\x95\x96\xb8\x74\x03\x00\x00xpw\x0c\x00\x00\x10?@\x00\x00\x00\x00\x02\x1e
\x01t\x00\x83eyJ0eXAiOiJKV1QiLCJhbGciOiJIUzI1NiJ9.eyJleHAiOjE2NTkzMjgyMTgsInVzZXJuYW1lIjoibGlmgmigfQ.sfS4we5cOGz9Ur258L
KT5cwKuhzYHwUiC6uKFKfaE0cxx\x00\x01q\x00\x00\x05x"
```

图 7-20　添加 Redis 缓存后的用户授权信息

再次发送同样的请求，观察 IDEA 的控制台输出，发现本次请求只执行了 doGetAuthenticationInfo 方法进行身份认证，而没有执行 doGetAuthorizationInfo 方法，如图 7-21 所示。授权信息是通过 Redis 获取的，而没有通过访问数据库获取，因此大大减少了和数据库的连接次数，提高了系统性能。

需要说明的是，每次请求都需要执行 CustomRealm 类中的 doGetAuthenticationInfo 方法，这是因为该方法要对 Token 值进行认证，包括用户名、密钥和 Token 是否过期等，并且此 Token 认证没有通过查询数据库进行验证，而是根据用户名（通过 token 获取）、密钥和系统当前时间生成一个新的 Token 值。授权信息相对来说比较固定，一个是和 @RequiresRoles 注解中的角色值匹配的用户角色，另一个是和 @RequiresPermissions 注解

```
 ─CustomRealm身份认证方法──
2022-08-19 20:54:45.434 DEBUG 12616 --- [io-8081-exec-10] o.s.web.servlet.DispatcherServlet
2022-08-19 20:54:45.435 DEBUG 12616 --- [io-8081-exec-10] s.w.s.m.m.a.RequestMappingHandlerMapping
2022-08-19 20:54:45.436 DEBUG 12616 --- [io-8081-exec-10] s.w.s.m.m.a.RequestMappingHandlerMapping
2022-08-19 20:54:45.437 DEBUG 12616 --- [io-8081-exec-10] o.s.web.servlet.DispatcherServlet
```

图 7-21 添加 redis 缓存后再次访问接口的 IDEA 控制台输出

的 value 值匹配的用户 user 表中的 permission 值及用户角色 role 表中的 permissions 值。这些授权信息在首次授权成功时写入 Redis，用户在请求其他 Controller 资源时可以从 Redis 中获取。

假如 token 值认证失败，doGetAuthenticationInfo 方法将会抛出 AuthenticationException 异常，程序也就不再执行权限认证。用户重新登录成功并获取新的 Token 值后，原来存入 Redis 的用户授权信息就不能再使用。

7.8 后端接口设计

根据平台总体规划和各功能模块的设计，需要设计以下接口：

(1) findRoutesByRole 接口。请求方式为 GET，参数为角色 role，拥有角色为系统管理员、学生个人、辅导员、系主任、学生处处长和教务处处长，返回值为 List < Routes > 类型对象，即路由元素的列表数据。

(2) findByName 接口。请求方式为 GET，参数为角色 username，拥有角色为系统管理员、学生个人、辅导员、系主任、学生处处长和教务处处长，返回值为 List < User > 类型对象，即用户对象元素的列表数据。

(3) commitgraphbyuser 接口。请求方式为 POST，参数为用户名 username 和个人电子签名 graph(Base64 编码格式)，拥有角色为系统管理员、学生个人、辅导员、系主任、学生处处长和教务处处长，返回值为 Map < String, Object > 类型对象，即键值对形式的对象。

7.8.1 findRoutesByRole 接口设计

根据平台总体设计，系统划分为系统管理员子系统、学生个人信息子系统、辅导员管理子系统、系主任管理子系统、学生处管理子系统和教务处管理子系统 6 个子系统，根据 6 个子系统需要分配 6 种角色作为系统参与者。因此，根据各角色的权限范畴，需要对每种角色进行页面菜单项的分配，包括子菜单的分配，以完成页面权限控制。要达到这一目的，需要数据库表角色表 role、路由表 routes 和子路由表 children 进行数据支持，3 个表的完整性约束如图 7-22 所示。

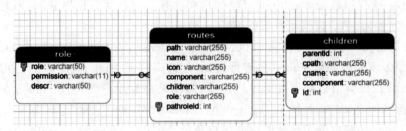

图 7-22 role、routes 和 children 数据完整性约束图

从数据完整性约束可以看出：路由表 routes 的字段 pathroleId 为主键，字段 role 为外键，依赖于角色表 role 的主键 role；子路由表 children 的字段 id 为主键，字段 parentId 为外键，依赖于 routes 的主键 pathroleId。

findRoutesByRole 接口设计的实现步骤如下。

1. 实现 Controller 控制层处理类

在 Demo 项目的 com.imooc.demo.web 包的用户管理类 AdminController 中，新建 findRoutesByRole 方法。

findRoutesByRole 方法文件代码

```
@GetMapping("/findRoutesByRole")
@RequiresPermissions("normal")
@RequiresRoles(logical = Logical.OR, value = { "user", "admin", "student", "depp", "stum", "dean" })
public Map<String, Object> findRoutesByRole(String role) {
    Map<String, Object> modelMap = new HashMap<String, Object>();
    List<Routes> routes = (List<Routes>) userService.findRoutesByRole(role);
    modelMap.put("routes", routes);
    return modelMap;
}
```

在上述代码中，@GetMapping("/findRoutesByRole")是对前端发起的/admin/findRoutesByRole 的 GET 请求进行拦截，请求接口的角色认证和权限认证与 findByUsername 接口相同。

该类中注入 UserService 接口对象 userService，通过调用 userService 的 findRoutesByRole 方法，根据前端传来的 role 参数查询出 Routes 类对象的 List 列表数据，再通过 resultMap 结果类封装后给前端。

2. 实现 Service 业务处理接口和 Service 接口实现类

（1）在 Demo 项目的 com.imooc.demo.service 包的 UserService 接口中，新建方法 findRoutesByRole，参数是由 Controller 层传来的用户名 role，返回值为 List<Routes> 对象。

UserService 接口的 findRoutesByRole 方法文件代码

```
public List<Routes> findRoutesByRole(String role);
```

（2）在 Demo 项目的 com.imooc.demo.service.impl 包的 UserServiceImpl 类中，创建 findRoutesByRole 方法，实现 UserService 接口。该实现类注入了 Mapper 接口的 UserMapper 对象，调用 findRoutesByRole 方法。同时该类注入了 RedisTemplate 接口的 redisTemplate 对象，其目的是从 Redis 中查询 key 为"menus_" + username 的角色路由数据，如果能查到，则返回角色路由数据；如果没有查到，则从数据库查找，查到后存入 Redis 数据库，返回角色路由数据。

UserServiceImpl 类的 findRoutesByRole 方法文件代码

```java
@Autowired
private UserMapper usermapper;
@Autowired
private RedisTemplate redisTemplate;
@Override
    public List<Routes> findRoutesByRole(String role, String username) {
        Object object = redisTemplate.opsForValue().get("menus_" + username);
        if (object != null) {
           return (List<Routes>) object;
        } else {
    List<Routes> objectm = (List<Routes>) usermapper.findRoutesByRole(role);
            if (objectm != null) {
                redisTemplate.opsForValue().set("menus_" + username, objectm);
                return objectm;
            } else {
                return null;
            }
        }
    }
```

3. 设计 Dao 接口和 XML 文件，实现对数据库的操作

（1）在 Demo 项目的 com.imooc.demo.dao 包的 UserMapper 接口中，创建方法 findRoutesByRole，参数是由 Service 层传来的用户名 role，返回值为 List<Routes>对象。

UserMapper 接口的 findRoutesByRole 方法文件代码

```java
List<Routes> findRoutesByRole(String role);
```

（2）在资源文件夹 resources 目录下的 userMapper.xml 配置文件中，创建 id 为 findRoutesByRole 的 SQL 语句，对数据库查询的结果标签类型为 resultMap，id 为 RoutesResult 的嵌套对象数据，实现 routes 和 children 表的联合查询。

findRoutesByRole 的 SQL 语句文件代码

```xml
<select id="findRoutesByRole" parameterType="String"
        resultMap="RoutesResult">
    SELECT r.*, c.cpath, c.cname, c.ccomponent, c.id, c.parentId
    FROM routes r left join children c on r.pathroleId = c.parentId
    where r.role = #{role}
</select>
```

4. /findRoutesByRole 请求的测试

打开 Postman 测试工具，在地址栏输入 127.0.0.1:8081/admin/findRoutesByRole?role=admin，请求方法选择 GET。另外，将用户登录成功获得的 token 值放入 Headers 中（key 为 token，value 为 token 值），单击 send 按钮，访问成功后就会获得 JSON 格式的角色权限路由数据。PostMan 测试接口获得路由数据如下所示，Redis 中存储的用户 liuzh 的角色路由数据如图 7-23 所示。

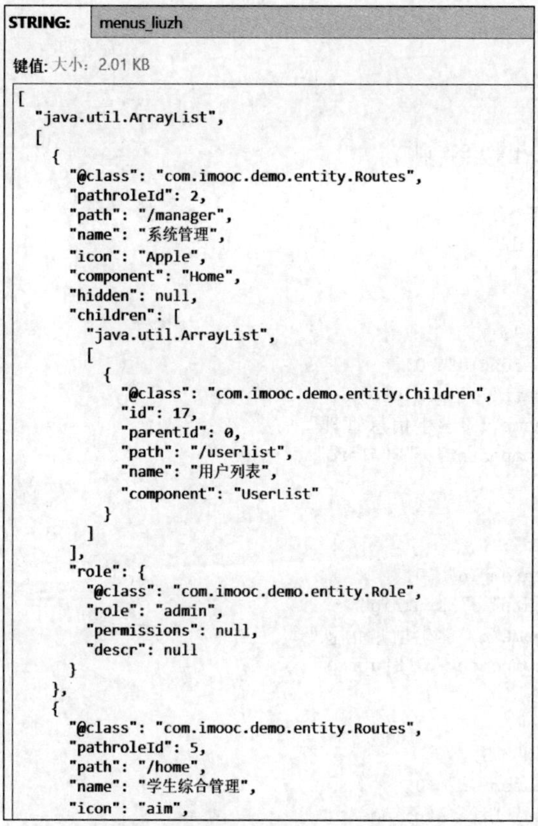

```
STRING:   menus_liuzh
键值: 大小: 2.01 KB
[
  "java.util.ArrayList",
  [
    {
      "@class": "com.imooc.demo.entity.Routes",
      "pathroleId": 2,
      "path": "/manager",
      "name": "系统管理",
      "icon": "Apple",
      "component": "Home",
      "hidden": null,
      "children": [
        "java.util.ArrayList",
        [
          {
            "@class": "com.imooc.demo.entity.Children",
            "id": 17,
            "parentId": 0,
            "path": "/userlist",
            "name": "用户列表",
            "component": "UserList"
          }
        ]
      ],
      "role": {
        "@class": "com.imooc.demo.entity.Role",
        "role": "admin",
        "permissions": null,
        "descr": null
      }
    },
    {
      "@class": "com.imooc.demo.entity.Routes",
      "pathroleId": 5,
      "path": "/home",
      "name": "学生综合管理",
      "icon": "aim",
```

图 7-23 Redis 中用户 liuzh 的角色路由数据

PostMan 测试接口获得路由数据

```
{
"routes": [
    {
        "pathroleId": 2,
        "path": "/manager",
        "name": "系统管理",
        "icon": "Apple",
        "component": "Home",
        "hidden": null,
        "children": [
            {
                "id": 17,
                "parentId": 0,
                "path": "/userlist",
                "name": "用户列表",
                "component": "UserList"
            }
        ],
        "role": {
            "role": "admin",
            "permissions": null,
            "descr": null
```

```json
            }
        },
        {
            "pathroleId": 5,
            "path": "/home",
            "name": "学生综合管理",
            "icon": "aim",
            "component": "Home",
            "hidden": null,
            "children": [
                {
                    "id": 3,
                    "parentId": 0,
                    "path": "/stuinfo",
                    "name": "学生信息管理",
                    "component": "StuInfo"
                },
                {
                    "id": 4,
                    "parentId": 0,
                    "path": "/stutrig",
                    "name": "学生增量信息",
                    "component": "Stutrig"
                },
                {
                    "id": 5,
                    "parentId": 0,
                    "path": "/stuimage",
                    "name": "学生个人信息",
                    "component": "Stuimage"
                },
                {
                    "id": 12,
                    "parentId": 0,
                    "path": "/stuInfoclass",
                    "name": "学生信息",
                    "component": "StuInfoClass"
                }
            ],
            "role": {
                "role": "admin",
                "permissions": null,
                "descr": null
            }
        },
        {
            "pathroleId": 12,
            "path": "/home",
            "name": "学籍异动查询",
            "icon": "edit",
            "component": "Home",
            "hidden": null,
            "children": [
                {
```

```json
            "id": 20,
            "parentId": 0,
            "path": "/mainAdmin",
            "name": "休学信息",
            "component": "mainAdmin"
        }
    ],
    "role": {
        "role": "admin",
        "permissions": null,
        "descr": null
    }
},
{
    "pathroleId": 16,
    "path": "/home",
    "name": "个人中心",
    "icon": "Apple",
    "component": "Home",
    "hidden": null,
    "children": [
        {
            "id": 23,
            "parentId": 0,
            "path": "/userInfo",
            "name": "用户信息",
            "component": "UserInfo"
        }
    ],
    "role": {
        "role": "admin",
        "permissions": null,
        "descr": null
    }
}
]
}
```

7.8.2 findByName 接口设计

findByName 接口主要实现用户信息的查询,用户调用此接口可查询用户自己的用户名、用户备注、角色、角色描述和电子签名等信息,系统将这些信息返回到前端显示。

findByName 接口设计的实现步骤如下。

1. 实现 Controller 控制层处理类

在 Demo 项目的 com.imooc.demo.web 包的用户管理类 AdminController 中,新建 findByName 方法。

findByName 方法文件代码

```
@GetMapping("/findByName")
@RequiresRoles(logical = Logical.OR, value = { "user", "admin", "student", "depp", "stum", "dean" })
```

```java
public Map<String, Object> findByName(String username) {
    System.out.println(username);
    Map<String, Object> modelMap = new HashMap<String, Object>();
    User user = userService.findByUsername(username);
    List<User> userList = new ArrayList<User>();
    userList.add(user);
    modelMap.put("userList", userList);
    return modelMap;
}
```

在上述代码中，@GetMapping("/findByName")是对前端发起的/admin/findByName 的 GET 请求进行拦截，角色和权限与 findRoutesByRole 接口设计相同。

通过调用 userService 的 findByUsername 方法，根据前端传来的 username 参数对 user 表和 role 表进行联合查询，再通过 Map<String，Object>对象 modelMap 将提交结果封装后给前端。

2．实现 Service 业务处理接口和 Service 接口实现类

（1）在 Demo 项目的 com.imooc.demo.service 包的 UserService 接口中，新建方法 findByUsername，参数是由 Controller 层传来的用户名 username，返回值为 User 对象。

UserService 接口的 findByUsername 方法文件代码

```java
public User findByUsername(String username);
```

（2）在 Demo 项目的 com.imooc.demo.service.impl 包的 UserServiceImpl 类中，创建 findByUsername 方法，实现 UserService 接口。该实现类注入了 Mapper 接口的 UserMapper 对象，调用 findByUsername 方法。

UserServiceImpl 类的 findByUsername 方法文件代码

```java
@Override
public int findByUsername (String username) {
    return usermapper.findByUsername(username);
}
```

3．设计 Dao 接口和 XML 文件，实现对数据库的操作

（1）在 Demo 项目的 com.imooc.demo.dao 包的 UserMapper 接口中，创建方法 findByUsername，参数是由 Service 层传来的用户名 username，返回值为 User 对象。

UserMapper 接口的 findByUsername 方法文件代码

```java
User findByUsername(String username);
```

（2）在资源文件夹 resources 目录下的 userMapper.xml 配置文件中，创建 id 为 findByUsername 的 SQL 语句，查询的结果标签类型为 resultMap，id 为 UserResult 的嵌套对象数据，实现 user 和 role 表的联合查询。

findByUsername 的 SQL 语句文件代码

```xml
<select id="findByUsername" parameterType="String"
        resultMap="UserResult">
    SELECT u.username,
           u.password,
```

```
                u.ban,
                u.permission,
                u.autograph,
                u.note,
                r.role,
                r.permissions,
                r.descr
        FROM user u left join role r on u.role = r.role
        where u.username = #{username}
</select>
```

7.8.3 commitgraphbyuser 接口设计

commitgraphbyuser 接口主要用于用户提交电子签名，根据审批流程和权限将电子签名镶嵌到异动审批表的 PDF 格式的电子文档中，代替以往人工的手写签字，实现远程无纸化办公。

commitgraphbyuser 接口设计的实现步骤如下。

1. 实现 Controller 控制层处理类

在 Demo 项目的 com.imooc.demo.web 包的用户管理类 AdminController 中，新建 commitGraphByUser 方法。

commitGraphByUser 方法文件代码

```java
@PostMapping("/commitgraphbyuser")
@RequiresRoles(logical = Logical.OR, value = { "user", "admin", "student", "depp", "stum", "dean" })
public Map<String, Object> commitGraphByUser(@RequestParam("username") String username,
        @RequestParam("graph") String graph) {
    Map<String, Object> modelMap = new HashMap<String, Object>();
    int rows = userService.commitGraphByUser(username, graph);
    if (rows > 0) {
        modelMap.put("msg", "照片提交成功!");
    } else {
        modelMap.put("msg", "照片提交失败!");
    }
    return modelMap;
}
```

在上述代码中，@PostMapping("/commitgraphbyuser")是对前端发起的/admin/commitgraphbyuser 的 POST 请求进行拦截，角色和权限与 findRoutesByRole 接口设计相同。

通过调用 userService 的 commitGraphByUser 方法，根据前端传来的 username 参数，对 user 表中的 autograph 字段用 Base64 编码的 graph 参数进行更新，再通过 Map<String, Object>对象 modelMap 将提交结果封装后给前端。

2. 实现 Service 业务处理接口和 Service 接口实现类

(1) 在 Demo 项目的 com.imooc.demo.service 包的 UserService 接口中，新建方法 commitGraphByUser，参数是由 Controller 层传来的用户名 username 和电子图像 graph，返回值为更新数据库表 user 影响的行数，返回值类型为 int。

UserService 接口的 commitGraphByUser 方法文件代码

```
public int commitGraphByUser(String username, String graph);
```

（2）在 Demo 项目中的 com.imooc.demo.service.impl 包的 UserServiceImpl 类中创建 commitGraphByUser 方法，实现 UserService 接口。该实现类注入了 Mapper 接口的 UserMapper 对象，调用 commitGraphByUser 方法。

UserServiceImpl 类的 commitGraphByUser 方法文件代码

```
@Override
public int commitGraphByUser(String username, String graph) {
    return usermapper.commitGraphByUser(username,graph);
}
```

3. 设计 DAO 接口和 XML 文件，实现对数据库的操作

（1）在 Demo 项目的 com.imooc.demo.dao 包的 UserMapper 接口中，创建方法 commitGraphByUser，参数是由 Service 层传来的用户名 username 和电子图像 graph，返回值为更新数据库表 user 影响的行数，返回值类型为 int。

UserMapper 接口的 commitGraphByUser 方法文件代码

```
int commitGraphByUser(String username, String graph);
```

（2）在资源文件夹 resources 目录下的 userMapper.xml 配置文件中，创建 id 为 commitGraphByUser 的 SQL 语句，标签为 <update>，实现对数据库表 user 的更新操作，更新的结果类型 int 可以省略。

findRoutesByRole 的 SQL 语句文件代码

```
<update id="commitGraphByUser" parameterType="String">
    UPDATE user
    SET autograph = #{graph}
    WHERE username = #{username}
</update>
```

7.8.4 用户退出登录 logout 接口设计

用户退出登录的设计思路是，首先通过 SecurityUtils.getSubject 方法获取当前用户主体 currentUser，再通过 currentUser.getPrincipal 方法获取 token 值，最后通过 JWTUtil.getUsername(token) 方法获取登录用户名 username。

（1）用 currentUser.logout 方法注销 Subject 后，该 Subject 的实例被再次引用将会被认为是匿名的。
（2）删除 Redis 中的 token 数据和用户角色路由数据。
（3）返回前端注销成功的提示。
（4）注销失败后返回前端注销失败的提示。

用户退出登录 logout 文件代码

```
@PostMapping(path = "/logout")
public Map<String, Object> logout() {
```

```java
        Map<String, Object> modelMap = new HashMap<String, Object>();
        Subject currentUser = SecurityUtils.getSubject();
        String token = (String) currentUser.getPrincipal();
        String username = JWTUtil.getUsername(token);
        try {
            currentUser.logout();
            redisTemplate.delete(username);
            redisTemplate.delete("menus_" + username);
            modelMap.put("msg", "注销成功!");
        } catch (AuthenticationException e) {
            e.printStackTrace();
            modelMap.put("msg", "注销失败!");
        }
        return modelMap;
    }
```

第8章

基于Spring Boot+Shiro+Vue开发的前后端分离学生信息管理项目整合实战——前端开发

本章学习目标
- 熟悉前端框架整合思路。
- 熟悉前端框架整合环境构建。
- 熟悉前端框架整合实现与测试。

通过前面的学习,读者掌握了将 Spring Boot 作为开发环境,整合 Shiro、JWT 和 MyBatis 开发后端的方法。本章介绍前端框架 Vue 的重要知识点,并将 VSCode 作为开发环境,对整合 Vue 开发前端框架进行详细的讲解。

8.1 开发思路整合

本章开发任务要求:
了解基于 VSCode 环境开发的基于 Vue 框架的学生信息管理前端项目。
本章主要内容:
(1) 前端框架的环境搭建;
(2) 前端项目的配置文件;
(3) 前端用户登录模块设计。

本系统前端框架整合了 Node.js、Element-plus、Axios、Vuex 等技术进行前端开发。虽然是前后端分离的架构,但是前端开发的每个请求都要与后端接口严密对接,无论是请求规则还是数据传递。所以描述前端离不开后端,当然开发后端也要充分考虑前端,只有这样才能开发出一套优秀的前后端分离的 Web 应用。

8.2 前端系统环境搭建

8.2.1 Vue 框架介绍

Vue 是一套用于构建用户界面的渐进式框架。Vue 采用自底向上增量开发的设计，提供了 MVVM 数据绑定和可组合的组件系统，具有简单、灵活的 API，通过简单的 API 可实现响应式的数据绑定和可组合的视图组件。

1. 主要特点

（1）轻量级。

Vue 能自动追踪依赖的模板表达式和计算属性，具有简单、灵活的 API，使开发人员容易理解和更快上手。

（2）双向数据绑定。

Vue 采用数据劫持结合发布者-订阅者模式的方式，通过 Object.defineProperty() 劫持各个属性的 set 和 get，在数据变动时发布消息给订阅者，触发相应的监听回调来渲染视图。

（3）指令。

Vue 与页面进行交互是通过内置指令完成的。指令的作用是当表达式的值改变时相应地将交互行为应用到 DOM 上。

（4）组件化。

Component 可以扩展 HTML 元素，封装可重用的代码。父子组件通信可通过 props 从父向子单向传递通信，子组件与父组件通过触发事件 $emit 通知父组件改变数据。兄弟组件通信有 Bus、Vuex 等；跨级组件通信则有 Bus、Vuex、provide/inject、$attrs/ $listeners 等。

（5）路由。

Vue-router 是 Vue 的路由插件，用于构建单页面应用。路由用于设定访问路径并将路径和组件映射起来，而传统的则是通过超链接实现页面的切换和跳转。

（6）状态管理。

Vuex 是一个专为 Vue 应用程序开发的状态管理模式。它其实就是一个单向的数据流，State 驱动 View 的渲染，而用户对 View 进行操作产生 Action，使 State 产生变化，从而使 View 重新渲染并形成一个单独的组件。

2. 主要优势

（1）Vue 组件化开发，减少代码量，易于理解。

（2）Vue 使用路由，不会刷新页面，只会局部刷新。

（3）Vue 使用 MVVM 开发模式，进行双向数据绑定。

（4）Vue 采用虚拟 DOM，可进行服务端渲染。

8.2.2 前端环境搭建

需要必要的开发环境，才能完成 Vue 3.0 项目搭建。项目需要安装的软件有 Node.js、

Visual Studio Code(简称 VS Code)。安装好 Node.js 后,再安装 Vue 3.0(vue-cli),这样 Vue 3.0 环境项目搭建就基本完成。

1. Node.js 的安装

Node.js 是一个基于 Chrome V8 引擎的 JavaScript 运行环境。Node.js 使用了一个事件驱动、非阻塞式 I/O 的模型,使其轻量又高效。npm(包管理工具)是基于 Node.js 的前端项目包管理工具,是项目中对各种程序包的依赖管理。传统的开发项目主要是后端,随着技术的更新,前端有了框架的开发模式管理,也需要用包管理工具的思想进行管理,目的是简化第三方程序包在项目中引用复杂化。

打开官网 https://nodejs.org/en/ 即可下载 Node.js,本项目使用的版本为 v17.7.1。Node.js 官网如图 8-1 所示。

图 8-1 node.js 官网

安装好 Node.js 后,在 cmd 中输入命令 node -v 即可查看安装的 Node.js 版本,这里显示版本为 v17.7.1 则说明 Node.js 安装成功,如图 8-2 所示。

2. 安装 Visual Studio Code

Visual Studio Code(简称 VS Code),是前端开发非常好用的一个代码编辑器。直接下载安装最新版即可,官网下载地址为 https://code.visualstudio.com,如图 8-3 所示。

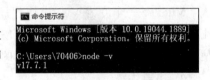

图 8-2 Node.js 安装成功

安装好的 VS Code 界面如图 8-4 所示。单击菜单"帮助",选择"关于"即可查看 VS Code 版本,本项目版本为 1.70.2(user setup),如图 8-5 所示。

3. 安装 Vue 3.0(Vue-cli)

安装好 Node.js 后就可以使用 npm(软件包管理工具)对项目所依赖的软件包进行下载及安装,npm 适合比较大型的应用。由于 npm 使用国外网站,使用起来比较慢,因此建议使用 cnpm 镜像进行安装,运行速度会快一些。打开 cmd,输入下列命令安装 cnpm:

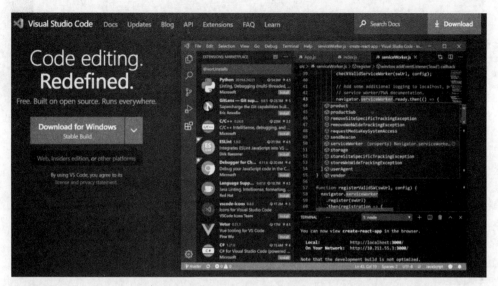

图 8-3　VS Code 官网

图 8-4　VS Code 界面

```
npm install -g cnpm --registry=https://registry.npm.taobao.org
```

在 cnpm 安装成功后，再用 cnpm 安装 Vue-cli，输入命令如下：

```
cnpm install --global vue-cli
```

打开 cmd，查看 Vue 版本，输入命令 vue -V，显示版本为@vue/cli 5.0.4，说明 Vue 安装成功，如图 8-6 所示。

图 8-5　VS Code 版本

图 8-6　Vue 安装成功

8.2.3　创建 Vue 3.0 项目

1. 创建一个 Vue 3.0 项目

开发环境搭建完成后，就可以创建一个 Vue 3.0 项目了。选择一个物理路径，如 D:/，在 cmd 中切换路径到 D 盘根目录，输入命令 vue create yeb 即可创建项目名称为 yeb 的 Vue 3.0 项目，默认使用 Vue 3.0 版本创建项目，如图 8-7 所示。

项目创建完成后，cmd 界面会出现"Successfully created project yeb."的提示，并且可以用 cd yeb 命令切换到项目路径下，通过 npm run serve 启动项目，如图 8-8 所示。

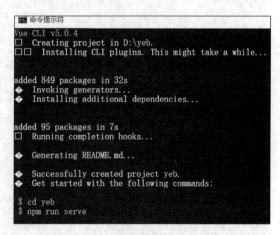

图 8-7　创建 Vue 3.0 项目界面　　　　　图 8-8　启动项目界面

项目启动成功后，出现以下提示则说明在本机的 8080 端口进行了发布，且局域网在 http://192.168.124.5:8080/地址进行了发布，如图 8-9 所示。

打开浏览器，在地址栏中输入 http://localhost:8080/，出现 Vue 欢迎页面，说明项目创建成功，如图 8-10 所示。

打开项目路径，查看项目结构，如图 8-11 所示。

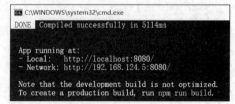

图 8-9　启动项目成功界面

图 8-10　Vue 欢迎页面

图 8-11　Vue 3.0 项目初始结构

2. 添加项目到 VS Code

打开 VS Code，在菜单"文件"中再选择"将文件夹添加到工作区"，选择项目 yeb 即可将项目添加到 VS Code。

8.2.4　项目目录结构

根据项目需要，在 src 目录下添加了 service、router、store 和 utils 子目录，yeb 项目在 VS Code 的目录结构如图 8-12 所示，并对项目结构的各个目录进行说明。

（1）node_modules：该目录为项目依赖文件。

（2）public：一般存放不会变动的文件（相当于 Vue-cli2.x 中的 static）。

（3）src：项目开发文件，包含许多子文件目录。

（4）src/assets：一般存放可能会变动的文件。

（5）src/components：自定义组件。

（6）src/router：存放项目所有 Vue 页面的路由配置。

（7）src/service：存放对 Axios 进行 interceptors 请求响应拦截的封装。

图 8-12　VS Code 中的 Vue 3.0 项目结构

（8）src/store：存放对状态管理 Vuex 的封装。

（9）src/utils：一是存放 api.js，其作用是在项目登录时对请求和响应进行拦截；二是存放 menus.js，其作用是对用户角色动态路由的数据获取。

8.3 前端项目的配置文件

8.3.1 package.json

package.json 定义了这个项目所需要的各种模块，以及项目的配置信息（如名称、版本、许可证等元数据）。

package.json 配置文件说明：

（1）创建项目时会自动生成 package.json 文件。npm install 命令根据该配置文件自动下载所需的模块，也就是配置项目所需的运行环境和开发环境。

（2）通过修改 package.json 文件的内容，再通过 npm install 进行更新。操作方法：切换路径到项目目录下，用命令 npm install 或 npm install --save-dev 进行安装，就会将 package.json 中的模块安装到 node-modules 文件夹下。

（3）package.json 中不能添加中文注释，否则会出现编译错误。

（4）package.json 文件可以手工编写，也可以使用 npm init 命令自动生成。

package.json 配置文件节点说明：

（1）name：项目名，即 version 项目版本。

（2）scripts：指定运行脚本命令的 npm 命令行缩写。例如，下列设置中指定了 npm run serve 所要执行的命令为 vue-cli-service serve（即 npm run＋快捷名）。

（3）dependencies：指定项目运行所依赖的模块。

（4）devDependencies：指定项目开发所需要的模块。

（5）eslintConfig：eslint 配置。eslint 工具是为了保证代码的一致性和避免一些语法错误。

（6）browserslist：用以兼容各种浏览器。

package.json 文件代码
——
```
{
 "name": "yeb",
 "version": "0.1.0",
 "private": true,
 "scripts": {
  "serve": "vue-cli-service serve",
  "build": "vue-cli-service build",
  "lint": "vue-cli-service lint"
 },
 "dependencies": {
  "@element-plus/icons-vue": "^1.1.4",
  "axios": "^0.26.1",
  "core-js": "^3.8.3",
  "element-plus": "^2.1.5",
```

```json
    "less-loader": "^10.2.0",
    "node-sass": "^7.0.1",
    "pdfjs-dist": "^2.5.207",
    "qs": "^6.10.3",
    "sass-loader": "^12.6.0",
    "style-loader": "^3.3.1",
    "vue": "^3.2.13",
    "vue-router": "^4.0.14",
    "vuex": "^4.0.2"
  },
  "devDependencies": {
    "@babel/core": "^7.12.16",
    "@babel/eslint-parser": "^7.12.16",
    "@vue/cli-plugin-babel": "~5.0.0",
    "@vue/cli-plugin-eslint": "~5.0.0",
    "@vue/cli-service": "~5.0.0",
    "eslint": "^7.32.0",
    "eslint-plugin-vue": "^8.0.3",
    "file-loader": "^6.2.0",
    "js-export-excel": "^1.1.4",
    "less": "^4.1.2",
    "less-loader": "^10.2.0",
    "node-sass": "^7.0.1",
    "pdfjs-dist": "^2.5.207",
    "sass-loader": "^12.6.0",
    "style-loader": "^3.3.1",
    "stylus-loader": "^7.0.0",
    "vue-pdf": "^4.3.0",
    "vue-pdf-embed": "^1.1.2",
    "vue-template-compiler": "^2.6.14"
  },
  "eslintConfig": {
    "root": true,
    "env": {
      "node": true
    },
      "extends": [
      "plugin:vue/vue3-essential",
      "eslint:recommended"
    ],
      "parserOptions": {
      "parser": "@babel/eslint-parser"
    },
    "rules": {}
  },
  "browserslist": [
    "> 1%",
    "last 2 versions",
    "not dead",
    "not ie 11"
  ]
}
```

8.3.2　App.vue

App.vue 是项目根组件，即页面入口文件，所有页面都是在 App.vue 下进行切换的，它负责构建定义及页面组件归集。App.vue 中的节点 <router-view> 一般用于路由管理，当需要对页面进行局部刷新时，就可以使用 <router-view>。在使用 <router-view> 渲染页面前，需要对 router.js 进行配置。<router-view> 具体使用场景：例如，创建一个侧边导航栏，并且需要使用侧边导航栏对主页面内容进行切换，主页面切换时保持侧边导航栏不变。

App.vue 文件代码

```
<template>
  <div id="app"></div>
  <router-view :key="$route.fullPath" />
</template>
<script>
export default {
  name: "App",
};
</script>
<style>
</style>
```

8.3.3　main.js

main.js 是程序主入文件，用于加载各种公共组件。

1. main.js 主要作用

（1）实例化 Vue。

（2）导入项目中经常会用到的插件和 CSS 样式。例如，Axios、Element-plus 等。

（3）定义全局变量。

2. 导航守卫 router.beforeEach

router.beforeEach 主要是通过跳转或取消的方式守卫导航，在前端路由跳转中或路由跳转前都会经过 beforeEach，而 beforeEach 可以通过 next 控制跳转哪个路由。每个守卫方法都有以下 3 个参数。

（1）to：进入哪个路由。

（2）from：从哪个路由离开。

（3）next：路由的控制参数，常用的有 next(true) 和 next(false)。

3. createApp(App)

使用 createApp 返回一个提供应用上下文的应用实例。可以在其后链式调用其他方法，主要有以下几种，如表 8-1 所示。

表 8-1　createApp 链式调用方法

参　　数	用　　法
component(组件)	注册或检索全局组件。注册还会使用给定的 name 参数自动设置组件的 name
use(使用)	安装 Vue.js 插件。在同一个插件上多次调用此方法时，该插件将仅安装一次

续表

参　数	用　法
config(配置)	包含应用配置的对象。同 Vue2 中 config,提供统一配置
directive(指令)	注册或检索全局指令。指令是一组具有生命周期的钩子
mixin(混入)	在整个应用范围内应用混入。一旦注册,它们就可以在当前应用中的任何组件模板内使用它。插件作者可以使用此方法将自定义行为注入组件,但不建议在应用代码中使用
mount(挂载)	应用实例的根组件挂载在提供的 DOM 元素上。同 Vue2 中的 el
provide(搭配 Inject)	设置一个可以被注入应用范围内所有组件中的值。组件应该使用 inject 来接收 provide 的值,provide 和 inject 绑定不是响应式的
unmount(卸载)	在提供的 DOM 元素上卸载应用实例的根组件

main.js 文件代码

```
import App from './App.vue'
import router from './router'
import service from "@/service/axios";
import ElementPLUS from 'element-plus'
import 'element-plus/theme-chalk/index.css'
import { createApp } from 'vue'
import qs from "qs";
import { initMenu } from './utils/menus'
import store from './store'
const app = createApp(App);
router.beforeEach((to, from, next) => {
    if (window.sessionStorage.getItem("tokenStr")) {
        initMenu(router, store);
        next();
    } else {
        if (to.path == '/') {
            next();
        }
    }
}),
//全局挂载 Axios 实例 service
app.config.globalProperties.$http = service
app.use(ElementPLUS)
    .use(store)
    .use(router)//路由进行挂载
    .mount('#app')
```

8.3.4　vue.config.js

vue.config.js 文件可以对项目进行 webpack 配置、插件及规则的配置、跨域请求的配置等,本项目中主要对跨域请求进行了相关配置。

在前后端分离的项目中,通常前端应用和后端 API 服务器没有运行在同一个主机上。在开发环境下将 API 请求代理到 API 服务器,这可以通过 vue.config.js 中的 devServer.proxy 选项配置,具体参数如下。

(1) devServer 是 Vue-cli 在开发环境时提供的一个代理服务器。

(2) proxy：代理的参数。target 为接口的域名；changeOrigin 的值设置为 true，意味着 host 设置成 target；pathRewrite 为路径改写规则，设置 pathRewrite 的作用是因为正确的接口路径是没有 /api 的，所以需要用 '^/api':'' 表示请求接口时去掉 api。

(3) '/api': {} 表示接口是 '/api' 开头的才用代理，所以请求接口就要写成 /api/xx/xx 形式，最后代理的路径就是 http://127.0.0.1:8081/xx/xx。

vue.config.js 文件代码

```
const { defineConfig } = require('@vue/cli-service')
const path = require('path');
module.exports = defineConfig({
  transpileDependencies: true,
  lintOnSave: false,
  devServer: {
    host: 'localhost',
    port: '8080',
    proxy: {
      '/api': {
          target: 'http://127.0.0.1:8081/',  //接口的域名
        //ws: true,
        changeOrigin: true,
        'pathRewrite': {                     //路径改写规则
            '^/api': ''                      //以//api/为开头的改写为''
        }
      },
    }
  },
  configureWebpack: {
    devtool: 'source-map'
  },
  //webpack 配置
  chainWebpack: config => {
      config.module
        .rule('pdf')
        .test(/\.pdf$/)
        .use('file')
        .loader('file-loader')
        .end();
  }
})
```

另外，configureWebpack 配置为 devtool: 'source-map'。在对项目进行打包时，经过一系列编译和转换，最终会形成生产环境的项目代码，并将此部署至线上。生产环境代码和打包后的代码千差万别，当构建后的代码报错时，如果没有 SourceMap，很难将错误溯源至源代码，这对开发很不友好。SourceMap 形成了源代码和构建后代码的映射关系，比如打包后产生的 .map 文件（eg: test.js.map）即为 SourceMap 文件。

8.4 前端用户登录模块设计

本节通过用户登录成功后根据角色权限获取动态路由的案例，说明基于 Vue 的前后端分离框架的前端用户登录最基本的逻辑设计和技术实现。根据用户登录逻辑设计用户登录获取角色动态路由流程图，如图 8-13 所示。

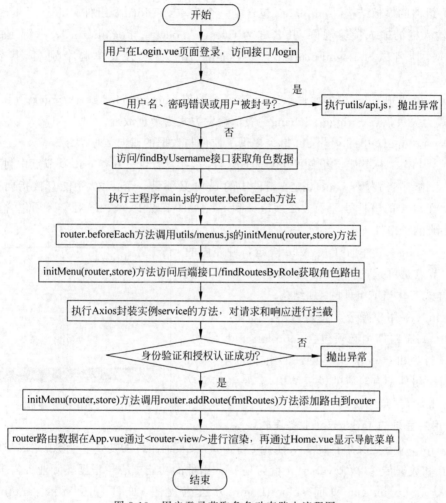

图 8-13 用户登录获取角色动态路由流程图

从该流程图中可以看出，登录模块共调用后端 3 个接口，分别是/login、admin/findByUsername 和 admin/findRoutesByRole，并通过 utils/api.js 和 service/axios.js 进行请求和响应的异常拦截。现在按逻辑走向对组件分别进行设计，一是 Login.vue（/login 接口和 findByUsername 接口访问）；二是 api.js；三是 main.js；四是 menus.js（initMenu 方法的 findRoutesByRole 接口访问）和 axios.js。

8.4.1 用户登录页面 Login.vue 设计

登录页面主要包含两个文本框和一个提交按钮，页面采用 ElementPlus 组件。文本框属性username、password用对象形式封装,变量名为LoginForm,用qs.stringify方法序列化后变成 username=liuzh&password=123456 形式,再作为参数提交给/login请求。<style>节点对登录表单LoginContainer、标题LoginTitle和记住我LoginRemember样式进行了外观和美化设计。

1. Vue的标准组成结构

Vue的标准组成结构分为<template>、<script>和<style>三部分。

(1) 组件的模板结构<template>包含了一个表单<el-form>组件,rules="rules"定义了表单输入框的输入检验规则,其名称为 rules;:model="LoginForm"表示绑定的表单输入数据,其名称为 LoginForm;ref="LoginForm"表示为表单起个别名,其名称为 LoginForm。

(2) 校验规则 rules 中的属性 trigger 有两个取值:blur 和 change。trigger:"blur"表示触发方式为失去焦点;trigger:"change"表示触发方式为数据改变。

(3) <el-input>中的 v-model 指令提供了控件与数据的双向数据绑定。

(4) <script>标签定义组件的 JavaScript 行为,开发者可以在<script>节点中封装组件的 JavaScript 业务逻辑。Vue 规定:组件中的 data 必须是一个函数,不能直接指向一个数据对象;组件中的事件处理函数必须定义到 methods 节点中;$refs 对象必须包含所有拥有 ref 注册的子组件。

(5) <style>组件的样式,Vue 规定:组件内的<style>节点是可选的,开发者可以在<style>节点中编写样式美化当前组件的 UI 结构。

Login.vue 用户登录页面如图 8-14 所示。

其中,每个组件中必须包含 template 模板结构,而 script 行为和 style 样式是可选的组成部分。Vue 规定:每个组件都有对应的模板结构,需要将其定义到<template>节点中。

图 8-14 用户登录页面

2. 用户登录的 JavaScript 业务逻辑

<script>标签定义组件的 JavaScript 行为,定义页面组件的变量和函数,完成与后端接口的交互和数据传递,故 JavaScript 行为是 Vue 组件的功能核心。根据实际业务需求,设计出用户登录 Login.vue 页面 submitLogin 方法流程图,如图 8-15 所示。methods 中 submitLogin 的方法代码见 Login.vue 文件代码。

基于 submitLogin 的方法,需要重点说明以下几点:

(1) 访问/login 接口用到的 Axios 插件需要在<script>标签中引入。

(2) this.$http 引用的是/service/axios.js 中对 Axios 封装的实例 service。

(3) Token、username、note、role.role、role.descr 等以键值对(key,value)的形式存储到 window.sessionStorage 中,便于前端页面使用,生命周期为关闭浏览器时失效。

(4) 在访问/api/admin/findByUsername 接口时,需要在 Headers 中携带用户登录成功后获取的 Token 值。

第8章 基于Spring Boot+Shiro+Vue开发的前后端分离学生信息管理项目整合实战——前端开发

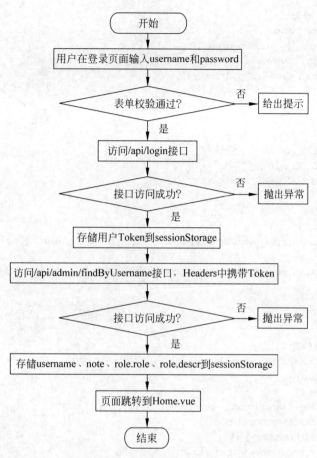

图 8-15 用户登录 Login.vue 页面 submitLogin 方法流程图

Login.vue 文件代码

```
<template>
  <div>
    <el-form :rules = "rules" ref = "LoginForm" :model = "LoginForm" class = "LoginContainer">
      <h3 class = "LoginTitle">系统登录</h3>
      <el-form-item prop = "username">
        <el-input type = "text" auto-complete = "false" v-model = "LoginForm.username" placeholder = "请输入用户名"></el-input>
      </el-form-item>
      <el-form-item prop = "password">
        <el-input type = "password" auto-complete = "false" v-model = "LoginForm.password" placeholder = "请输入密码"></el-input>
      </el-form-item>
      <el-checkbox v-model = "checked" class = "LoginRemember">记住我</el-checkbox>
      <el-button type = "primary" style = "width: 100%" @click = "submitLogin">登录</el-button>
    </el-form>
  </div>
</template>
<script>
import { postRequest } from "@/utils/api";
```

```javascript
import qs from "qs";
import axios from "axios";
import router from "@/router/index";
import store from "../store";
import { initMenu } from "../utils/menus";
export default {
  name: "Login",
  data() {
    return {
      LoginForm: {
      username: "liuzh",
      password: "123456",
      },
      rules: {
      username: [
        { required: true, message: "请输入用户名", trigger: "blur" },
      ],
      password: [{ required: true, message: "请输入密码", trigger: "blur" }],
      },
    };
  },
  methods: {
    submitLogin() {
      this.$refs.LoginForm.validate((valid) => {
        if (valid) {
          axios
            .post("/api/login", qs.stringify(this.LoginForm))
            .then((response) => {
              if (response) {
                const tokenStr = response.message;
                //存储用户token
                window.sessionStorage.setItem("tokenStr", tokenStr);
                console.log("login");
                this.$http
                  .get("/api/admin/findByUsername", {
                    params: {
                      username: this.LoginForm.username,
                    },
                    headers: {
                      token: window.sessionStorage.getItem("tokenStr"),
                    },
                  })
                  .then((res) => {
                    if (res) {
                      //存储用户user等
              window.sessionStorage.setItem("user", res.message.username);
              window.sessionStorage.setItem("note", res.message.note);
              window.sessionStorage.setItem("role", res.message.role.role);
              window.sessionStorage.setItem("rnote", res.message.role.descr);
                      router.push("/home");
                    }
                  });
              }
            })
```

```
          .catch(function (err) {
            if (err) {
            }
          });
      } else {
        this.$message.error("请输入所有字段!");
        return false;
      }
    });
  },
 },
};
</script>
<style>
.LoginContainer {
  border-radius: 15px;
  background-clip: border-box;
  margin: 180px auto;
  width: 350px;
  padding: 15px 35px 15px 35px;
  background: #fff;
  border: 1px solid #eaeaea;
  box-shadow: 8 8 25px #cac6c6;
}
.LoginTitle {
  margin: 8px auto 48px auto;
  text-align: center;
}
.LoginRemember {
  text-align: left;
  margin: 8px 8px 15px 8px; }
</style>
```

8.4.2 用户登录拦截器 api.js 设计

用户登录拦截器 api.js 主要作用是在用户登录请求/login 接口时，对用户请求 axios.interceptors.request 和用户响应 axios.interceptors.response 进行拦截。

在子目录 utils 中，新建 api.js 封装请求拦截器和响应拦截器，对用户登录请求/login 接口进行请求拦截和异常捕获，并由前端显示给用户，具体实现步骤如下。

(1) 请求拦截器 service.interceptors.request.use 中的 config 代码块是对响应成功的信息进行捕获，并对回调数据 config 进行返回。判断 window.sessionStorage 中是否包含 Token 信息，如果存在 Token，将 Token 以键值对的形式存储到 Headers 中。

(2) 响应拦截器 service.interceptors.response.use 中的 success 代码块是对响应成功的异常进行捕获和显示，并用 return success.data 返回回调数据。对 success.data.code 的值进行判断，该值等于 400，为用户名错误；该值等于 401，为密码错误；该值等于 402，为用户被封号。

(3) 响应拦截器 service.interceptors.response.use 中的 error 代码块是对响应失败的异常进行捕获和显示，用语句 ElMessage.error(error.message)对异常数据进行显示。如果服务器遇到错误，无法完成请求(Request failed with status code 500)等情况，则由

ElMessage.error(error.message)弹窗显示给用户。

（4）拦截器各代码块是串联式执行，注意各拦截器都要返回回调数据，否则访问成功后程序不能向下进行。

api.js 文件代码

```
import axios from 'axios'
import Message, { ElMessage } from 'element-plus'
import router from '../router'
//请求拦截器
axios.interceptors.request.use(config => {
    if (window.sessionStorage.getItem("tokenStr")) {
        config.headers['token'] = window.sessionStorage.getItem("tokenStr");
    }
    return config;
})
//响应拦截器
axios.interceptors.response.use(success => {
    //业务逻辑错误
    if (success) {
        if (success.data.code == 400 || success.data.code == 401 || success.data.code == 403 || success.data.code == 402|| success.data.code == 408) {
            ElMessage.error(success.data.message);
        }
        if (success.data.message) {
            //ElMessage.success(success.data); //页面弹出提示
        }
    }
    return success.data;
},
    error => {
        ElMessage.error(error.message);
        return Promise.reject(error);
    }
);
```

8.4.3 用户请求 Controller 接口的 axios.js 设计

为了使用户在访问任何一个接口时，对系统异常都有一个准确的捕获并显示给用户，现设计一个封装 axios.js 的实例 service。

在子目录 service 中，通过 axios.create 方法生成一个名为 service 的 Axios 实例，封装请求拦截器和响应拦截器，在访问每个 Controller 接口时进行请求拦截和异常捕获，尤其对验证 Token、捕获后端的全局异常等起到重要的作用，并由前端显示给用户。封装好的 Axios 实例由语句 export default service 完成属性暴露。

（1）请求拦截器 service.interceptors.request.use 中的 config 代码块是对响应成功的信息进行捕获，并对回调数据 config 进行返回。

（2）响应拦截器 service.interceptors.response.use 中的 success 代码块是对响应成功的异常进行捕获和显示，判断 success.data.code 的值是否等于 408，如果为真则为 Token 验证错误，最后用 return success.data 返回回调数据。

(3) 响应拦截器 service.interceptors.response.use 中的 error 代码块是对响应失败的异常进行捕获和显示,并由路由 router.push("/unauthorized/")转发至 unauthorized.vue 页面,该页面 intPage 方法请求后端接口/api/unauthorized,获取后端 CustomRealm 类的 doGetAuthenticationInfo 方法中身份认证时抛出的异常。如果有 token 认证失败、用户不存在和该用户被封号等情况,则由 unauthorized.vue 页面显示给用户,最后用 return Promise.reject(error)返回回调数据。

(4) 拦截器各代码块是串联式执行,注意各拦截器都要返回回调数据,否则访问成功后程序不能向下进行。

(5) 该封装实例在 main.js 中设置了全局挂载,代码为 app.config.globalProperties.$http=service,供所有页面使用。

axios.js 文件代码

```
import axios from 'axios'
import Message, { ElMessage } from 'element-plus'
import router from "@/router/index";
/****** 创建 Axios 实例 ******/
const service = axios.create({
    timeout: 10000  //请求超时时间
});
service.defaults.headers.post["Content-Type"] = "application/x-www-form-urlencoded";
service.defaults.withCredentials = true;  //让 AJAX 携带 Cookie
service.defaults.headers["Content-Type"] = "application/x-www-form-urlencoded";
//请求拦截器
service.interceptors.request.use(config => {
    if(config){
     console.log(config)
    }
    if (window.sessionStorage.getItem("tokenStr")) {
        config.headers['token'] = window.sessionStorage.getItem("tokenStr");
    }
return config;
})
//响应拦截器
service.interceptors.response.use(success => {
       if (success) {
         if (success.data.code == 408 ) {
            ElMessage.error(success.data.message);
         }
         if (success.data.message) {
            //ElMessage.success(success.data);  //页面弹出提示
         }
       }
    return success.data;
    },
    error => {
        {
         router.push("/unauthorized/");
         //对响应错误进行操作
         return Promise.reject(error);
        }
```

```
      });
export default service;
```

8.4.4 获取动态路由 menus.js 设计

根据后端 findRoutesByRole 接口设计,请求该接口需要传递参数 role 和 Headers 中携带的 Token,访问成功后回调数据应为根据角色 role 查询出的路由表 routes 和子路由表 children 的动态路由数据。该路由数据要经过格式化处理,并与 router/index.js 中的静态路由相匹配,才能根据角色动态分配角色路由,生成角色权限的页面菜单项,达到角色页面权限划分的目的,并且不用创建多个角色页面。

设计思路:在 menus.js 中设计 initMenu 方法,访问/api/admin/findRoutesByRole 接口。

获取动态路由 menus.js 设计的实现步骤如下。

(1) 设计对/api/admin/findRoutesByRole 接口的访问,通过参数 role 查询路由表 routes、子路由表 children 中该角色 role 所拥有的路由和子路由数据 dataMenu(实参)。

(2) 通过方法 formatRoutes(routes)(routes 为形参)对路由数据进行格式化,即通过 formatRoutes 方法将服务器返回的角色路由的 JSON 格式数据转为 router 需要的格式。首先定义路由所需属性 path、component、name、icon、children,然后通过 routes.forEach 对后端查询出的路由的元素进行遍历,添加到以各属性值为 key 的 value 值中。其中对组件变量 component 进行了解析,让组件名 component 指向/views/目录下的组件名 component (变量组件名).vue,并对子路由 children 格式化进行递归调用。最后形成以属性为 key、数据库查询数据为 value 的键值对元素的一维数组数据 fmtRoutes。

```
component(resolve) {
        require(['../views/' + component + '.vue'], resolve);
    }
```

(3) 将一维数组数据 fmtRoutes 添加到路由 router,再通过 Vue.app 的<router-view>标签渲染到 Home.vue 中进行页面显示。

(4) 在 router/index.js 中,设计所有的静态路由结构,包括主路由、子路由和对 Vue 组件的加载等,最后通过语句 export default router 进行属性暴露。

menus.js 文件代码

```
import service from "@/service/axios";
export const initMenu = (router, store) => {
    if (store.state.routes.length > 0) {
        return;
    }
    service.get("/api/admin/findRoutesByRole", {
        params: {
            role: window.sessionStorage.getItem("role"),
        },
        headers: {
            token: window.sessionStorage.getItem("tokenStr"),
        }
    }).then(resp => {
```

```
            if (resp) {
                let dataMenu = [];
                if(resp.routes){
                    dataMenu = resp.routes
                }else{
                    dataMenu = resp.data.routes
                }
                let fmtRoutes = formatRoutes(dataMenu);
                //添加到路由
                router.addRoute(fmtRoutes);
                //数据存入Vuex
                store.commit('initRouts', fmtRoutes)
            }
        })
    }
export const formatRoutes = (routes) => {
    let fmtRoutes = [];
    routes.forEach(router => {
        let {
            path,
            component,
            name,
            icon,
            children
        } = router;
        if (children && children instanceof Array) {
            //递归
            children = formatRoutes(children);
        }
        let fmtRouter = {
            path: path,
            name: name,
            icon: icon,
            children: children,
            component(resolve) {
                require(['../views/' + component + '.vue'], resolve);
            }
        }
        fmtRoutes.push(fmtRouter)
    });
    return fmtRoutes;
}
```

router/index.js 文件代码

```
import { createRouter, createWebHashHistory } from 'vue-router'
import Stutrig from '../views/Stutrig.vue'
import StuInfo from '../views/StuInfo.vue'
import Stuimage from '../views/StuImage.vue'
import Home from '../views/Home.vue'
import mainAdmin from '../views/mainAdmin.vue'
import mainDepp from '../views/mainDepp.vue'
import PdfView from '../views/PdfView.vue'
import StuInfoClass from '../views/StuInfoClass.vue'
import mainStum from '../views/mainStum.vue'
```

```javascript
import mainDean from '../views/mainDean.vue'
import mainSoun from '../views/mainSoun.vue'
import UserList from '../views/UserList.vue'
import StuImageOnly from '../views/StuImageOnly.vue'
import UserInfo from '../views/UserInfo.vue'
import Unauthorized from '../views/unauthorized.vue'

const routes = [
    {
        path: '/',
        name: '用户登录',
        component: () => import('../views/Login.vue'),
        hidden: true,
        disable: true
    },
    {
        path: '/manager',
        name: '系统管理',
        icon: 'aim',
        component: Home,
        hidden: false,
        children: [
            {
                path: '/userlist',
                name: '用户列表',
                component: UserList,
            },
        ]
    },
    {
        path: '/home',
        name: '学生综合管理',
        icon: 'edit',
        component: Home,
        hidden: false,
        children: [
            {
                path: '/stuinfo',
                name: '学生信息管理',
                component: StuInfo
            },
            {
                path: '/stutrig',
                name: '学生增量信息',
                component: Stutrig
            },
            {
                path: '/stuimage',
                name: '学生个人信息',
                component: Stuimage
            },
            {
                path: '/stuInfoclass',
                name: '学生信息',
```

```
                    component: StuInfoClass
                },
            ],
        },
        {
            path: '/home',
            mode: "history",
            name: '审批流程管理',
            icon: 'aim',
            component: Home,
            hidden: false,
            children: [
                {
                    path: '/mainSoun',
                    name: '辅导员审批',
                    component: mainSoun,
                },
                {
                    path: '/mainDepp',
                    name: '系主任审批',
                    component: mainDepp,
                },
                {
                    path: '/pdfView',
                    name: 'pdf预览',
                    component: PdfView,
                },
                {
                    path: '/mainStum',
                    name: '学生处审批',
                    component: mainStum,
                },
                {
                    path: '/mainDean',
                    name: '教务处审批',
                    component: mainDean,
                },
                {
                    path: '/mainAdmin',
                    name: '休学信息',
                    component: mainAdmin,
                },

            ]
        },
        {
            path: '/home',
            name: '个人中心',
            icon: 'aim',
            component: Home,
            hidden: false,
            children: [
                {
                    path: '/userInfo',
```

```
                    name: '用户信息',
                    component: UserInfo,
                },
            ]
        },
        {
            path: '/home',
            name: '个人信息查询',
            icon: 'aim',
            component: Home,
            hidden: false,
            children: [
                {
                    path: '/stuImageOnly',
                    name: '学籍信息',
                    component: StuImageOnly
                },
            ]
        },
        {
            path: '/unauthorized',
            name: 'shiro 错误信息',
            icon: 'aim',
            component: Unauthorized,
            hidden: false,
        },
    ]
    const router = createRouter({
        history: createWebHashHistory(),
        routes
    })
    export default router;
```

8.4.5 用户登录成功页面显示

（1）系统管理员角色登录成功页面显示。

系统管理员角色具有系统管理、学生综合管理、学籍异动管理和个人中心这 4 个主菜单项，系统管理主菜单下有用户列表子菜单项，学生综合管理主菜单下有学生信息管理、学生增量信息、学生个人信息和学生信息这 4 个子菜单项，学籍异动查询主菜单下有休学信息子菜单项，个人中心主菜单下有用户信息子菜单项。系统管理员角色登录成功页面如图 8-16 所示。

打开浏览器，按 F12 键，打开浏览器开发工具，切换到"网络"菜单界面，重新刷新登录成功页面，可以看到出现/findRoutesByRole?role＝admin 的一条新请求，右侧"预览"中会显示通过接口查询后端数据库中的角色 role＝admin 的动态路由数据 routes，如图 8-17 所示。

（2）辅导员角色登录成功页面显示。

辅导员角色具有学生综合管理、学籍异动管理和个人中心这 3 个主菜单项，学生综合管理主菜单下有学生信息子菜单项，学籍异动查询主菜单下有辅导员审批子菜单项，个人中心主菜单下有用户信息子菜单项。辅导员角色登录成功页面如图 8-18 所示。

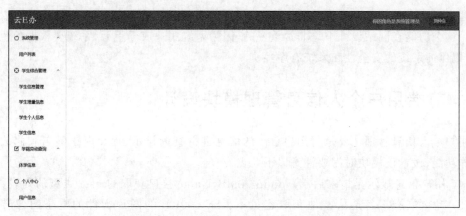

图 8-16　系统管理员角色登录成功页面

图 8-17　浏览器控制台中角色 admin 的动态路由数据 routes

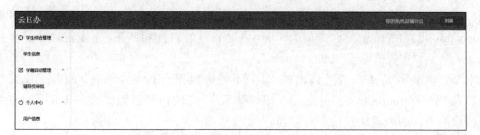

图 8-18　辅导员角色登录成功页面

打开浏览器，按 F12 键，打开浏览器开发工具，切换到"网络"菜单界面，重新刷新登录成功页面，可以看到出现/findRoutesByRole？role＝user 的一条新请求，右侧"预览"中会显示通过接口查询后端数据库中的角色 user 的动态路由数据 routes，如图 8-19 所示。

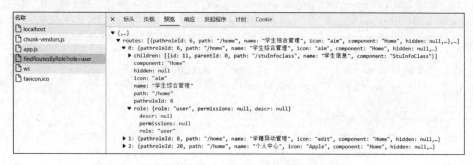

图 8-19　浏览器控制台中角色 user 的动态路由数据 routes

通过以上两种角色登录成功的案例，实现了前后端分离项目根据角色划分页面权限的功能。这个功能在网站设计中是非常重要的，因为系统用户成千上万，系统要根据角色划分权限，以保证系统的安全性和系统有序性。

8.5 前端用户个人信息管理模块设计

用户个人信息管理主要是对用户自己的信息进行查询显示、提交电子签名和修改密码等操作功能，设计该模块时需要注意以下几点。

（1）用户信息显示是请求后端/admin/findByName接口获取的user类型的列表数据。

（2）提交电子签名是从本地获取容量不大于500KB的图片文件的电子图像转换成Base64编码数据后，通过/admin/commitgraphbyuser提交给后端。

（3）修改密码是在输入正确的原密码的前提下设置新的密码，密码通过MD5加密。

（4）用户信息页面<template>主要包含显示用户信息和更新密码的隐式表单。

8.5.1 用户信息页面 UserInfo.vue 设计

显示用户信息主要使用 Element-plus 的<el-table>控件显示用户信息，用户信息源用：data="userList"进行数据绑定，userList 是请求 admin/findByName 接口成功后返回的用户列表数据。

（1）<template>视图设计。

在<template>标签下添加<el-table>标签，再添加<el-table-column>子标签，用来显示用户属性值。标签中的 prop 为 userList 的属性名，该属性名要与后端访问接口 findByName 得到的 userList（类型为 List<User>）的属性相匹配，才能正确地显示用户信息。

电子签名显示是从后端查询出的 Base64 编码数据，用<el-image>的:src 属性指向 Base64 编码数据 scope.row.autograph 即可显示。:src 的值是图像文件的 URL，也就是引用该图像的文件的绝对路径或相对路径。

UserInfo.vue 视图文件代码
```
    <div>
      <h3>用户信息</h3>
</div>
<el-table :data="userList" border stripe style="width: 100%">
  <el-table-column prop="username" label="用户名" width="128" />
  <el-table-column prop="note" label="用户备注" width="128" />
  <el-table-column prop="role.role" label="角色" width="100" />
  <el-table-column prop="role.descr" label="角色描述" width="150" />
  <el-table-column prop="autograph" label="电子签名" width="150">
      <!-- 图片的显示 -->
      <template v-slot="scope">
        <el-image style="width: 50px; height: 50px" :src="scope.row.autograph" />
      </template>
  </el-table-column>
```

```
        <el-table-column label="上传照片" width="230">
          <template v-slot="scope" #dropdown>
            <input class="input_image left" type="file" @change="uploadImage($event, scope.row.username)"
              accept="image/*" />
          </template>
        </el-table-column>
        <el-table-column label="操作" width="100">
          <template v-slot="scope" #dropdown>
            <el-button type="primary" size="mini" @click="shareu(scope.row.username)">修改密码</el-button>
          </template>
        </el-table-column>
</el-table>
```

(2) JavaScript 业务逻辑。

在 JavaScript 代码段的 methods 中新建 initStuInfo 方法,请求接口为/admin/findByName,请求方式为 GET,参数为 username。请求成功后,resp.data.userList 响应数据被变量 this.userList 接收,进而在视图中显示用户信息。另外,为了让用户切换到 UserInfo.vue 页面就能显示用户信息,initStuInfo 方法需要在 mounted 方法中挂载。

initStuInfo 方法文件代码
——
```
initStuInfo() {
    this.$http
      .get("/api/admin/findByName", {
        params: {
          username: window.sessionStorage.getItem("user"),
        },
        headers: {
          token: window.sessionStorage.getItem("tokenStr"),
        },
      })
      .then((resp) => {
        if (resp.data) {
          this.userList = resp.data.userList;
        } else {
          this.userList = resp.userList;
        }
      });
},
```
initStuInfo 方法挂载
——
```
mounted() {
  this.initStuInfo();
}
```

(3) 用户信息页面显示。

用户登录成功后,单击左侧导航条"个人中心",单击"用户信息"即可显示用户个人信息。用户信息页面如图 8-20 所示。

图 8-20 用户信息页面

8.5.2 更新密码的隐式表单设计

（1）<template>视图设计。

在 UserInfo.vue 视图设计中添加<el-table-column>标签，设置 label="操作"，再添加<el-button>标签，设置触发事件为 shareu，并携带行参数 scope.row.username，其作用是将隐式更新密码表单显示。

更新密码的隐式表单使用 Element-plus 的<el-form>控件，外层<div>节点使用 v-show 的值 popupu 为 false 或 true 设置表单的显示或隐藏，popupu 值默认为 false。

更新密码视图代码

```
<!-- 更新密码 -->
  <div class="footer" :class="[popupu ? 'popyes' : 'popno']" :model="updateForm" ref="updateFormRef" label-width="100px"
    style="width: 311px; height: 360px; background-color: bisque" v-show="popupu == true">
    <p class="top">确认更新密码吗?</p>
    <div class="bot" v-show="popupu == true">
      <el-form :rules="addFormRules">
        <el-form-item label="用户名" prop="username">
<el-input v-model="updateForm.username" :data="updateForm.username" disabled="false"></el-input>
        </el-form-item>
        <el-form-item label="原密码" prop="oldpassword"><el-input v-model="updateForm.oldpassword" show-password></el-input>
        </el-form-item>
        <el-form-item label="新密码" prop="newpassword"><el-input v-model="updateForm.newpassword" show-password></el-input>
        </el-form-item>
        <el-form-item label="确认新密码" prop="confirmpassword"><el-input v-model="updateForm.confirmpassword" show-password></el-input>
        </el-form-item>
      </el-form>
      <p class="bot-rig" @click="commitu">确认</p>
      <p class="bot-lef" @click="cancelu">取消</p>
    </div>
  </div>
```

（2）JavaScript 业务逻辑。

在 JavaScript 代码段的 methods 中新建 shareu 方法，请求接口为/admin/findByUsername，请求方式为 GET，参数为 username。请求成功后，用 this.updateForm.username 变量接收响应数据，目的是在隐式弹窗的用户名文本框中显示当前用户名。

在 methods 中新建 commitu 方法，请求接口为/ login 和/admin/updatePassword，请求接口为/ login 的目的是验证用户名和现密码的正确性，请求接口为/admin/updatePassword 目的是根据新修改密码进行 user 表的密码更新。另外，前端对输入的新密码进行两次输入的校验，输入不一致时给出提示，密码为空时也给出提示。

Shareu 方法文件代码

```
shareu(id) {
    this.popupu = true;
        this.$http({
        method: "GET",
        url: "/api/admin/findByUsername",
        params: {
          username: id,
        },
        headers: {
          token: window.sessionStorage.getItem("tokenStr"),
        },
    }).then((res) => {
      if (res) {
        this.updateForm.username = res.message.username;
      }
    });
},
```

commitu 方法文件代码

```
commitu() {
    if (
      !this.updateForm.oldpassword ||
      !this.updateForm.newpassword ||
      !this.updateForm.confirmpassword
    ) {
      alert("密码不能为空");
      return;
    }
    if (
      this.updateForm.newpassword != this.updateForm.confirmpassword
    ) {
      alert("两次输入的新密码不一致");
      return;
    }
    this.$http({
      method: "POST",
      url: "/api/login",
      data: qs.stringify({
        username: this.updateForm.username,
        password: this.updateForm.oldpassword,
      }),
    }).then((res) => {
      if (res) {
        if (res.code == 200) {
            this.$http({
            method: "POST",
```

```
                    url: "/api/admin/updatePassword",
                      data: qs.stringify({
                      username: this.updateForm.username,
                      newpassword: this.updateForm.newpassword,
                    }),
                    headers: {
                      token: window.sessionStorage.getItem("tokenStr"),
                    },
                  }).then((result) = > {
                    if (result) {
                      alert(result.msg);
                      window.location.reload();
                      this.popupu = false;
                    }
                  });
                } else {
                  alert("原密码错误");
                }
              } else {
                alert("异常错误");
              }
            });
          },
```

（3）更新密码页面显示。

单击用户信息的"修改密码"按钮，弹出更新密码弹窗，如图 8-21 所示。

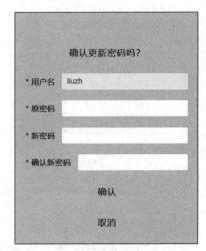

图 8-21　更新密码弹窗

当输入原密码错误时，弹出"原密码错误"的对话框，如图 8-22 所示。

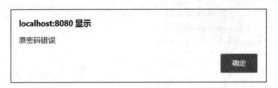

图 8-22　原密码错误弹窗

当输入新密码输入两次不一致时，弹出"两次输入的新密码不一致"的对话框，如图 8-23 所示。

当输入密码均输入正确时，弹出"密码修改成功"的对话框，如图 8-24 所示。

图 8-23　两次输入的新密码不一致弹窗　　　　图 8-24　密码修改成功弹窗

8.5.3　提交电子签名表单设计

（1）< template >视图设计。

在用户信息浏览数据行添加"上传照片"字段，添加< input >控件，type 为 file，@change

触发事件为 uploadImage($event, scope.row.username)，accept="image/*"规定能够上传的文件类型为图片文件。

```
<el-table-column label="上传照片" width="230">
    <template v-slot="scope" #dropdown>
    <input class="input_image left" type="file"
      @change="uploadImage($event, scope.row.username)"
      accept="image/*" />
```

（2）JavaScript 业务逻辑。

首先通过 e.target.files[0]获取图片文件，然后再判断文件是否大于 500KB，如果为真则给出提示；然后 new FileReader()生成一个对象 img，通过 img.readAsDataURL(file)将图片文件转化为 Base64 编码数据；最后通过访问/admin/commitgraphbyuser 接口，把 Base64 编码的图片数据上传给 user 表的 autograph 字段。上传成功后给出提示，并用 window.location.reload()语句刷新页面，显示上传的电子签名。

```
uploadImage(e, username) {
    //上传图片并预览
    let file = e.target.files[0]; //获取第一个文件
    const isLt500KB = file.size / 1024 / 1024 < 0.48;
    if (!isLt500KB) {
      this.$message.error("上传图片大小不能超过 500KB!");
      return;
    }
    let img = new FileReader();
    img.readAsDataURL(file); //将 img 转化为 Base64 编码数据
    img.onload = ({ target }) => {
      this.imgSrc = target.result;
      this.$http({
        method: "POST",
        url: "/api/admin/commitgraphbyuser",
        data: qs.stringify({
          username: username,
          graph: this.imgSrc,
        }),
        headers: {
          token: window.sessionStorage.getItem("tokenStr"),
        },
      }).then((result) => {
        if (result) {
          alert(result.msg);
          window.location.reload();
        }
      });
    };
},
```

（3）提交电子签名页面显示。

单击页面的"选择文件"按钮，选择图片文件后打开，即可上传电子图片，选择文件界面如图 8-25 所示。

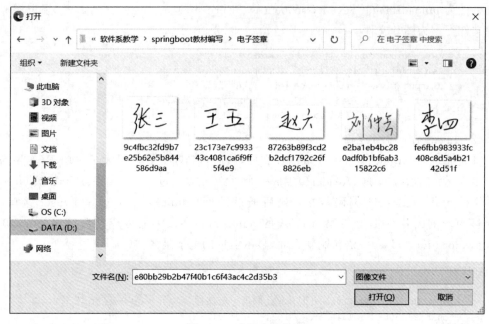

图 8-25 选择文件界面

当上传图片容量大于 500KB 给出提示,如图 8-26 所示。

当上传图片成功后给出提示,并可在页面浏览上传的电子签名,如图 8-27、图 8-28 所示。

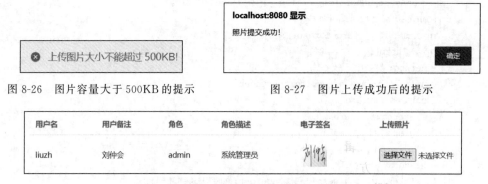

图 8-26 图片容量大于 500KB 的提示   图 8-27 图片上传成功后的提示

图 8-28 图片上传成功后电子签名的显示

8.6 前端用户注销登录模块设计

在用户使用系统过程中,为防止系统被他人非法操作,在不使用系统时注销用户是非常必要的。因为前后端分离项目不使用 Session,所以注销用户方法与前后端一体项目注销用户方法是不同的。前后端分离项目注销用户设计思路如下:

(1) 通过 Home.vue 页面的下拉列表框选择"注销登录"选项。

(2) 触发 commandHandler(cmd)事件,请求/logout 接口,Headers 中携带 Token。

(3) 访问成功后,后端将用户 Subject 注销,并将 Redis 中的 Token 和角色路由数据删除。

（4）前端将 sessionStorage 中存储的 tokenStr、user、note、role 和 rnote 移除，并将状态管理 Vuex 中的数据 initRouts 清空。

（5）获取后端 msg 弹窗显示给用户。

（6）页面跳转至登录页面 Login.vue。

（7）注销失败时获取后端 msg 弹窗显示给用户。

1. <template>视图设计

在<template>节点上添加<el-dropdown>，绑定事件@command="commandHandler"，添加<el-dropdown-menu>节点和<el-dropdown-item>子节点，指定方法 command="logout"，具体代码如下所示。

Home.vue 中的注销登录文件代码

```
<el-header class="homeHeader">
    <div class="title">云E办
    </div>
    <div class="annone">你的角色是{{ rnote }}</div>
    <el-dropdown class="userInfo" @command="commandHandler">
        <el-button type="primary">
            {{ note }}<el-icon class="el-icon--right">
                <arrow-down />
            </el-icon>
        </el-button>
        <template #dropdown>
            <el-dropdown-menu>
                <el-dropdown-item>个人中心</el-dropdown-item>
                <el-dropdown-item>设置</el-dropdown-item>
                <el-dropdown-item command="logout" divided>注销登录</el-dropdown-item>
            </el-dropdown-menu>
        </template>
    </el-dropdown>
</el-header>
```

2. JavaScript 业务逻辑

commandHandler 方法的指定参数 cmd 为 logout，指用户单击下拉列表的"注销登录"时触发的事件。为了防止用户操作失误，需要设计一个确认弹窗，询问用户是否确认退出登录，如果是，则执行 service 插件的 POST 方法。在请求接口/logout 时，注意 Headers 中是否携带 Token。请求成功后，清除 sessionStorage 中的信息和 Vuex 中的路由数据，并弹窗提示用户，最后页面跳转至登录页 Login.vue。

注销登录 js 文件代码

```
commandHandler(cmd) {   //该方法有一个参数,cmd
    if (cmd == 'logout') {
        this.$confirm("是否确认退出登录吗?", "提示", {
            iconClass: "el-icon-question",     //自定义图标样式
            confirmButtonText: "确认",          //确认按钮文字更换
            cancelButtonText: "取消",           //取消按钮文字更换
            showClose: true,                   //是否显示右上角关闭按钮
```

```
              type: "warning",                  //提示类型 success/info/warning/error
            }).then(function () {
              service({
                method: "POST",
                url: "/api/logout",
                headers: {
                  token: window.sessionStorage.getItem("tokenStr"),
                },
              }).then(res => {
                if (res) {
                  window.sessionStorage.removeItem("tokenStr")
                  window.sessionStorage.removeItem("user")
                  window.sessionStorage.removeItem("note")
                  window.sessionStorage.removeItem("role")
                  window.sessionStorage.removeItem("rnote")
                  store.commit('initRouts', [])
                  alert(res.msg)
                  router.push("/");
                }
              }).catch(function (err) {
                alert(err.msg)
              }
              )
            })
          }
        }
```

3. 注销用户页面显示

用户单击下拉列表的"注销登录",弹出退出登录的提示框,如图 8-29 所示。单击确认后,弹出注销成功的提示框,如图 8-30 所示,单击确认页面跳转至登录页面。查看 Redis 数据库可知,用户 liuzh 的 Token 及用户角色路由数据已被注销,浏览器控制台应用程序的会话存储中也没有了 sessionStorage 的相关信息。

图 8-29　注销登录的提示框　　　　　　图 8-30　注销成功的提示框

参 考 文 献

[1] 陈恒,楼偶俊,巩庆志,等.Spring Boot 从入门到实战(微课视频版)[M].北京：清华大学出版社,2020.
[2] 贾志杰.Vue+Spring Boot 前后端分离开发实战[M].北京：清华大学出版社,2021.
[3] 黑马程序员.Spring Boot 企业级开发教程[M].北京：人民邮电出版社,2019.
[4] 疯狂软件.Spring Boot2 企业应用实战[M].北京：电子工业出版社,2018.
[5] 王松.Spring Boot+Vue 全栈开发实战[M].北京：清华大学出版社,2019.

图书资源支持

感谢您一直以来对清华版图书的支持和爱护。为了配合本书的使用,本书提供配套的资源,有需求的读者请扫描下方的"书圈"微信公众号二维码,在图书专区下载,也可以拨打电话或发送电子邮件咨询。

如果您在使用本书的过程中遇到了什么问题,或者有相关图书出版计划,也请您发邮件告诉我们,以便我们更好地为您服务。

我们的联系方式:

清华大学出版社计算机与信息分社网站: https://www.shuimushuhui.com/

地　　址: 北京市海淀区双清路学研大厦 A 座 714

邮　　编: 100084

电　　话: 010-83470236　　010-83470237

客服邮箱: 2301891038@qq.com

QQ: 2301891038(请写明您的单位和姓名)

资源下载: 关注公众号"书圈"下载配套资源。

书 圈

清华计算机学堂

观看课程直播